Modeling by Object-Driven Linear Elemental Relations: A User's Guide for MODLER(c)

OPERATIONS RESEARCH/COMPUTER SCIENCE INTERFACE SERIES

Ramesh Sharda, Editor
Conoco/DuPont Chair of Management of Technology
Professor of Management Science & Information Systems
College of Business Administration
Oklahoma State University
Stillwater, Oklahoma, U.S.A.

Modeling by Object-Driven

Linear Elemental Relations:

A User's Guide for MODLER[c]

HARVEY J. GREENBERG
Mathematics Department, Campus Box 170
University of Colorado at Denver
P.O. Box 173364
Denver, CO 80217-3364
BITNET: hgreenberg@cudenver

Springer-Science+Business Media, B.V.

Library of Congress Cataloging-in-Publication Data

Greenberg, Harvey J.
Modeling by object-driven linear elemental relations : a user's guide for MODLER / Harvey J. Greenberg.
p. cm. -- (Operations research/computer science interface series)
"September 1992."
Includes index.
Additional material to this book can be downloaded from http://extras.springer.com.
ISBN 978-1-4613-6388-0 ISBN 978-1-4615-3168-5 (eBook)
DOI 10.1007/978-1-4615-3168-5
1. Linear programming--Data processing. 2. MODLER (Computer program language) 3. Operations research. I. Title. II. Series.
T57.74.G728 1993
658.4'033--dc20 92-43259
CIP

Printed on acid-free paper.

Acknowledgements

I wish to thank our industrial sponsors – Amoco Oil Company, IBM, Shell Development Company, Chesapeake Decisions, Inc., GAMS Development Corp., Ketron Management Science and MathPro Inc. Milton M. Gutterman (Amoco) provided some code and testing for some non-DOS environments; Thomas Dehne and Chris Strauss (Ketron) and David Heltne (Shell) provided testing for the CMS environment. Optimizers were provided by Roy Marsten (OB1) and Michael Saunders (MINOS). John Stone (Ketron) provided support for MINOS. I thank my colleague, Frederic H. Murphy, who inspired some of the recent enhancements and reviewed an earlier draft of this User's Guide.

I am grateful to the students who provided detailed criticisms of an earlier draft: Kenneth Cortes, Margaret Cortes, Dee Davis, Anastasia Grandbois, Charles Halsey, and Michael O'Brien.

Last, but not least, earlier versions of this work was partially supported by the Office of Naval Research (contract #N-00014-88-K-0104).

Contents

Introduction

Welcome to MODLER, which stands for *M*odeling by *O*bject-*D*riven *L*inear *E*lemental *R*elations. It is a language for representing linear programming models, completely separate from instances defined by data realizations. It also includes representations of binary variables and logical constraints, which arise naturally in large-scale planning and operational decision support. The basic input to MODLER is a model file, and its basic output is a matrix file that is in a standard (MPS) format for most optimizers and for ANALYZE and RANDMOD. MODLER can also generate a syntax file for ANALYZE to enable automatic translation of activities and constraints into English for intelligent analysis support.

Chapter 1 gives an overview of MODLER and how to install it in different computing environments. It also describes how to get started and how to obtain further documentation and on-line help.

Chapter 2 reviews the forms of linear programming models and describes the *syntax* of a model. One of the functions provided by MODLER is to enable simple expressions of a model in algebraic form. One can view the model, however, in block schema form as well as in its original algebraic form.

Chapter 3 first illustrates simple modeling expressions and on-line queries. Then, one of the randomization capabilities is described, which enables instances to be generated without actual data. The purpose in doing so is either for rapid prototyping, to see model properties for instances, or for experimentation with algorithms.

Chapter 4 begins with some precepts to distinguish models from instances that are defined by particular data. Then, special MODLER features are illustrated with more examples. The last section describes additional ways to randomize the assignments of values to data objects.

Chapter 5 progresses to some advanced uses of MODLER, starting with incorporation of logical constraints and binary decision variables into a linear programming framework. Then, an inventory problem is presented to show how to use control tables to express dynamic equations. The next section describes how to represent special nonlinear forms, which determine supply and demand quantities as functions of prices, but with increasing and decreasing returns to scale, respectively. The last section illustrates elastic programming representations with MODLER.

Some advice from a student

> If you are just learning linear programming, this User's Guide will introduce some concepts that require a second reading. This is especially true beginning with chapter 5, and even summations formalized in chapter 6 can be confusing at first.

Chapter 6 introduces the SUM expression. This is needed for two reasons. First, the expressive power of implied summations is limited without resorting to control tables that could distract from a natural formulation. Second, it documents the algebraic operations in the model's equations in a way that helps some to understand the model more quickly.

Chapter 7 describes dependency relations on data objects: parameters, sets and tables. First, expressions are defined that allow sets to be defined in terms of parameters and other sets. Second, table expressions are defined that allow tables to be defined in terms of parameters, sets and other tables. Finally, a command is described that enables views of these and other dependency relations in the model.

Chapter 8 describes environment controls, such as the use of the _SETUP command. It proceeds to explain the use of various file types that provide interfaces with other systems, such as optimizers and ANALYZE.

Chapter 9 goes further into illustrating the use of MODLER for model description and instance generation. It draws from previous chapters, and it compares different expressions of the same model. It proceeds to describe additional ways to obtain model documentation and manage scenarios.

Chapter 10 considers model management in the context of using MODLER and the companion system, ANALYZE. After giving an overview of model management, a case study is described. Finally, examination of directory files that came with MODLER is suggested for getting modeling assistance.

Appendices give more guidance to information about model formulation with MODLER. Appendix A reflects some experiences with MODLER, notably by students, who wished there was some guide to troubleshooting. Appendix B gives an overview of the MODLER library, which contains elementary, generic models, as well as models taken from the operations research literature.

Chapter 1. Overview, Installation, and Help

I have never had good relations with mechanical objects...I have a suntan lamp. As I sit under it, it rains on me.

Woody Allen

This chapter is designed for a quick entry into MODLER. The purpose of MODLER is to represent a model and generate instances of it. Some underlying questions are: What is a model? What roles do data play in understanding the modeling process? It will take many examples and discussions to answer these questions (among others).

One thing we can say is that MODLER is a software system, designed to accept representations of models and instances, defined by data realizations. It produces files for optimizers to solve instances and for other software systems, like ANALYZE, to provide computer-assisted analysis of results. In addition to file output, MODLER has extensive interactive query capability. One overview, that of information flow, is given in figure 1-1.

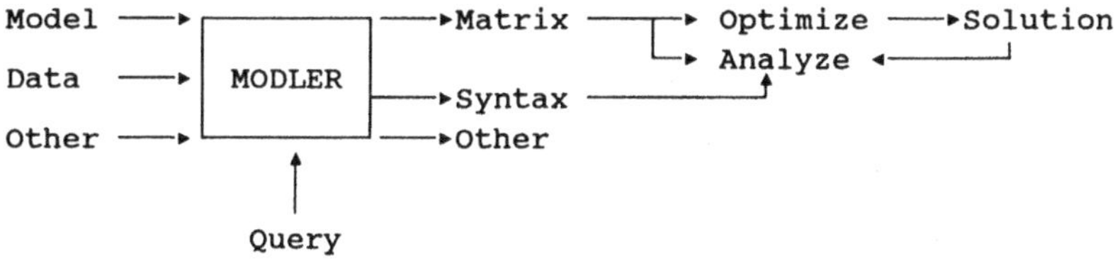

Figure 1-1. An Overview of MODLER

Installation instructions are given in the next section for a DOS environment. Instructions for installation in other environments are given in the subsequent section. Both installations are automated, but you may want to make further changes, such as the dimensions of the arrays. This is described in § 1.3.

Section 4 introduces some of the conventions in the interactive language of MODLER, such as abbreviations and some of the more popular environmental controls. Finally, the last section tells you how to obtain on-line help and further documentation.

1.1 What you have and how to begin.

You received one 3½" disk, created with DOS 3.3, with the MODLER files. Codes were compiled with Lahey(R) F77L-EM/32 5.0 and linked under DOS 3.3 (the executable uses the Phar Lap extender). Begin by copying the original disk for a backup. To install MODLER on your IBM PS2 (or a clone) under DOS, you will need about 1mb on your hard disk.

To run MODLER you need 2mb available memory. You should also set files and buffers to at least 20 in your CONFIG.SYS. A co-processor is not required.

If you have the full version, and if you want to run MODLER in another computing environment, see §1.3. You must first prepare the source codes, which contain INCLUDE statements. This has been automated with the utility in AUTOEDIT.ARC. All of this can be done automatically under DOS. The source code and data files will use a little less than 1mb on your hard disk. At 4800 baud it takes about 2 hours to upload all source code, data and documentation files.

This User's Guide is designed to be read sequentially. A reference manual can be generated by MODLER by specifying EXEC DOCMODL at the MODLER prompt. This and other information about on-line help is given in §1.5 (see, also, Chapter 8).

1.2 Quick installation for DOS environment.

To install in a DOS environment, put the MODLER disk into drive a: and change directory to the path where you want to install MODLER. It is recommended this be an empty directory, perhaps called MODLER. Then, at the DOS prompt, enter:

> COPY A:INSTALL.BAT [*drive*]

The default disk drive that you are using is a:, and you can change this by entering another drive. For example, enter INSTALL B to install from drive b (note no colon after B in the drive spec). The appropriate files will be extracted from the ARC files, and you will be ready to run MODLER.

Once the installation is complete, you may delete INSTALL.BAT from your hard disk. Your next step is to enter MODLER at the DOS prompt. After some sign-on information is displayed, you will be ready to test that you received everything and that all is well. After the MODLER prompt, enter EXEC ATEST. You will see a lot of messages flashing over your screen. These are from test commands being executed. After a few minutes, you should see a message that MODLER appears to be working, and you are left with the MODLER prompt. You may then continue with some experiments, such as those in this User's Guide. To leave MODLER, enter QUIT.

If you have Microsoft Windows 3.0, you should look at MODLER.PIF and MODL.BAT. You probably need to modify them for your environment.

1.3 Alternative environments.

If you have the full version of MODLER, you can prepare a load module for alternative environments. Besides DOS, MODLER has been tested on a VAX 8700/VMS and an IBM RS 6000. It is written in FORTRAN/77, so in principle it runs in any environment, but there are at least two things to consider. One is the file protocol, the other is the INCLUDE syntax for the compiler. For those reasons, and for increasing the dimensions to accommodate larger problems, the source codes must be edited before uploading them to a mainframe.

Note to VM/CMS users See NOTES.CMS for notes about upload, compilation and linking.

The preparation is automated in SETMODL.BAT, but you may need to modify some things if your environment is not explicitly recognized. To use the automatic preparation, create a directory on your hard disk and make sure it is empty (so you can upload all the files that will appear).

Insert the MODLER disk in your a: drive. At the DOS prompt, enter

> A:PKXARC A:MODLSUBS SETMODL.BAT

Then, simply enter

> SETMODL *environment* [*drive*]

where *environment* is one of: XT, VAX, TSO, CMS, XNX, PRIME, AIX or SUN4 (must be given in CAPS). This will extract source and data files for upload. The main program is in FLIPMAIN.

The default drive is a. If the floppy containing MODLER is not in your a drive, you must specify the drive. For example, SETMODL VAX B performs the setup for the VAX environment from the floppy in drive B (note: do not put the colon after the drive).

1.4 Some interactive language conventions.

Examples and documentation use a standard metalanguage with the following symbols.

All commands and keywords are upper case. Lower case words describe a type of specification.

Square brackets [] enclose an optional specification.

Curly braces {} enclose a list of options, one of which must be specified.

An ellipsis ... denotes continued repetition of the last option.

A vertical bar | denotes exclusive options.

Example 1: `SWITCH [name [...]]`

This specifies the language syntax for the SWITCH command. The outer square brackets denote that no specification need follow the SWITCH command (in which case the switches and their values are listed). If a specification does follow, it is the name of a switch (in which case the value of the switch is toggled between ON and OFF). The inner square brackets with the ellipsis means that additional switch names can be specified. The following illustrates the possibilities.

`SWITCH`

...lists switches and their settings

`SWITCH ECHO`

...changes value of switch named ECHO

`SWITCH ECHO ERROR`

...changes values of two switches, named ECHO and ERROR

Example 2: `DISPLAY {object} [name mask] [//options]`

```
where object := PARAMETR, SET, TABLE, ACTIVITY, BINARY,
                EQUATION, LOGICAL
```

This specifies the language syntax for the DISPLAY command. The Curly brackets indicate that some object must follow the DISPLAY specification. Since no vertical bar separates the object list, more than one can be specified. The following are thus valid specifications (their meanings will be described later).

```
DISPLAY PARAMETR
DISPLAY SET
DISPLAY PARAMETR, SET
DISPLAY SET, PARAMETR, TABLE, EQUATION
```

Note from the last specification that the objects can be specified in any order. Whenever a list allows any or all members to be specified, the asterisk can be used to specify all of them. Thus,

```
DISPLAY *
```

is equivalent to specifying all seven objects.

The first pair of square brackets indicates that a name mask can be specified (as before, the meaning of this will be defined later; here only the metalanguage is illustrated). The last pair of square brackets indicate an optional specification of more options that must be preceded by the double slash (//). For the DISPLAY command, the last set of options are:

```
{BRIEF | EXTENDED}, ATTRIBUT = [attribute]
```

The Curly bracket and vertical bar indicate that either BRIEF or EXTENDED is a valid specification, but both cannot be specified. The following are valid specifications are valid.

```
DISPLAY SET //BRIEF
DISPLAY SET //EXTENDED
DISPLAY SET //EXTENDED ATTRIBUT = PLACE
```

Example 3: `READIN {{MODEL | PACKED | FORMAT} filespec`
`| DATA [filespec]}`

This specifies the language syntax for the READIN command. The Curly braces indicate that one of the options must be specified, and the vertical bars indicate that exactly one must be specified. The square brackets indicate that the file specification (filespec) is optional (the default is described later). This means the following specifications are valid.

```
READIN MODEL  MYMODEL
READIN DATA   MYDATA
READIN PACKED MYFILE
READIN FORMAT MYFILE2
READIN DATA
```

Note that the `READIN DATA` does not require a file specification, but the others do.

As we go through the examples in the next chapter, you will experience some conventions about the interactive language (FLIP). One is that you must LOCK CAPS, because the interactive command language is case-sensitive, and all commands and keywords use capital letters.

Telescopic abbreviations are allowed as long as there is no ambiguity. For example, the telescopic abbreviations of `DISPLAY` are:

```
D
DI
DIS
DISP
```

```
DISPL
DISPLA
```

Since no other command begins with D, this is an allowed abbreviation. On the other hand, the READIN and RETURN commands both begin with RE, so at least 3 keystrokes are required for each of those commands.

Another convention is allowing *string variables*, which are local to the session, to reduce the number of keystrokes for repeated specifications. For example, one may define the string R as:

```
STRING R = READIN MODEL
```

then use 'R whenever you want to read in a model file. The single quote (') is the reserved character for denoting a string variable.

The STRING command can also be used to clear a string by specifying STRING *name*. To clear all strings, specify the asterisk (*) for the *name*.

It is also useful to change the width and length of your screen. This is done with the SCREEN command. The default (upon installation) sets WIDTH=79 and LENGTH=23. You can see your settings anytime by entering SCREEN (or simply SCR) at the MODLER prompt. To change either or both of these simply enter the desired number(s), such as SCR L=9999 to make the length virtually infinite. The screen setting is also used when spooling output to a file to determine page ejection (see OUTPUT command).

The strings, screen and other settings can be set in your PROFILE, which is an execution file that contains commands to be executed before beginning the interactive session. This is described in Chapter 8.

Additional information is available on-line (see next section). Especially look at the help for each command, and PRINT QUICKREF.DOC for a quick reference to all commands (the .DOC suffix is not needed if you are printing from within MODLER).

1.5 On-line documentation.

MODLER has a variety of facilities for providing on-line documentation. At the sign-on (or anytime) you may enter ? to obtain a general introduction. The list of commands is shown by entering HELP to the MODLER prompt. For information about any particular command, enter the command after HELP. For information about all commands, such as when generating your own file for printing, enter HELP *.

> It is recommended you try the HELP command
> (after EXEC ATEST is successful).

There is a dictionary that will help guide you to the nomenclature and other documentation sources. Enter ? DIR at the MODLER prompt to see what directory files are available, which you may PRINT. In particular, note the DIRDOCS dictionary entry, which guides you to some documentation files that you could PRINT for more explanations. (See §8.2 to create your own dictionaries.)

It is recommended that you enter EXEC DOCMODL at the MODLER prompt. This will generate a file, called MODLER.DOC, that contains most of the help and documentation available on-line. You might also examine the file DOCMODL.EXC (by printing it), which generates the MODLER.DOC file. This will illustrate how you might make your own, personalized documentation.

Each PRINT uses the screen width and length settings, even when the output is to a file (specified by the OUTPUT command). When you EXEC DOCMODL, you will see the screen settings and a prompt is given (from the _PAUSE command in DOCMODL.EXC) to see if you want to continue. If you forgot to increase the screen LENGTH, enter N to abort the execution of the documentation and specify SCREEN L=*length* to set the screen length to some integer (*length*), like 60, depending on your printer (see §8.3 for how to have this set with a PROFILE execution file before beginning your session).

When printing to a file, as above, page ejection (from the _EJECT command) is merely the FORTRAN 1H1 format. The output file, therefore, has a line with a 1 in column 1 wherever there should be a new page, which is determined by the SCREEN LENGTH setting.

Once you become familiar with how MODLER works, you may add your own documentation for others to have. Dictionaries and documentation files are two ways to supplement the on-line documentation, and the DOCMODL execution file is one way to obtain hard copy. After reading Chapter 8, you will know more ways to make MODLER suit your environment.

Chapter 2. Anatomy of a Linear Program

A theory has only the alternative of being right or wrong. A model has a third possibility—it may be right but irrelevant.

Manfred Eigen

In this chapter we consider the parts that comprise a linear program and how to define them. We begin with the algebraic form, which views the model as a collection of variables related by a system of equations and inequalities with an objective to minimize or maximize some linear form. Beyond the algebra, linear programs have composition rules that comprise its syntax.

2.1 Algebraic form.

One way to express an LP model, which is the way of most textbooks, is first to define sets, domains, data tables and variables, and second to define the constraints. To illustrate, here is a production-distribution model.

Given: 1. A collection of plants distinguished by their location and processes of operation. Each process requires raw material input and produces finished products.
2. Markets, distinguished by location and product.
3. Transportation links from plants to markets.

Data: a_{ipj} = amount of raw material i used by process p at plant j;

α_{pjk} = yield of product k from process p at plant j;

S_i = total supply of raw material i;

c_{pj} = unit operation cost of process p at plant j;

C_j = capacity limit of plant j, measured in terms of its total output;

s_{jmk} = unit shipping cost of product k from plant j to market m;

D_{mk} = demand for product k required to be satisfied in market m.

Conditions: If there is no link from plant j to market m for product k, the value of $s_{jmk} = \infty$.

Model: Let P_{pj} = level of production using process p at plant j;

Let T_{jmk} = amount of product k sent from plant j to market m.

Let COST = total cost of production and transportation.

Minimize COST: $P_{pj}, T_{jmk} \geq 0$;

$$\begin{array}{llll}
\text{COST} & = \Sigma_{pj}\, c_{pj}\, P_{pj} + \Sigma_{jmk}\, s_{jmk}\, T_{jmk} & & \\
\text{A(i)} & = \Sigma_{pj}\, a_{ipj}\, P_{pj} & \leq S_i & \text{(Raw material availabilities)} \\
\text{C(j)} & = \Sigma_p\, P_{pj} & \leq C_j & \text{(Capacity limits)} \\
\text{B(j,k)} & = \Sigma_p\, \alpha_{pjk}\, P_{pj} - \Sigma_m\, T_{jmk} & = 0 & \text{(Balance equations)} \\
\text{D(m,k)} & = \Sigma_j\, T_{jmk} & \geq D_{mk} & \text{(Demands)}
\end{array}$$

Notice that we began our expression with *Let ...* When we express the *model,* we write the objective and the constraints. This is the *algebraic form*. Alternatively, we could express the model in *schema form*, as shown in figure 2-1.

	P(p,j)	T(j,m,k)	
COST	PCOST	SHIPCOST	...MIN
A(i)	CONSUME		<= SUPPLY
B(j,k)	YIELD	-1	= 0
C(j)	1		<= CAPACITY
D(m,k)		1	>= DEMAND

Figure 2-1. A Schema of the Production-Distribution Model

The rows have been classified with COST the objective row, and the other row types begin with A, B, C and D, respectively. The activities have been classified as Production and Transportation. Tables PCOST, SHIPCOST, CONSUME, SUPPLY, YIELD, CAPACITY and DEMAND had to be declared, and they correspond, respectively, to the given c, s, a, S, α, C, and D. The subscripts in the original problem definition become *domains* in this expression (and in algebraic modeling languages). Each domain is a cross product of sets. The sets are:

i = raw material;
p = process;
j = plant (location);
k = finished product;
m = market.

When forming the name of a row or column in the schema, its type is distinguished by its first character. Then, its domain is used (without the parentheses and commas). For example, consider the transportation activity T(j,m,k) for the particular plant j=DE (a code for Denver), the particular market m=CH (a code for Chicago), and the particular finished product k=TB (a code for table). Then, the column name is TDECHTB. This syntax will be formally described in the next section.

2.2 Model syntax.

The *syntax* of a model is its rules of composition. This was introduced in the previous section and illustrated with a production-distribution model. When we form a matrix for linear program, the name of a row or column can contain information about its meaning. MODLER uses this in forming names, which can be passed to ANALYZE to provide syntactic translations of rows and columns in the matrix defined by the instance of the linear program.

The most elementary object in the syntactic description is the *set*. In MODLER, a set can be *explicit* (sometimes called *primitive*), or it can be *implicit*. An explicit set is declared simply by its announcement; an implicit set is specified with an equal sign (=) followed by the nature of its dependency on other sets or parameters. An explicit set's members simply comprise a list; an implicit set's members are determined by the syntax of its definition.

MODLER currently supports two types of implicit sets: unions of sets, previously defined, and numeric sets defined by consecutive integers, such as time periods. The syntax for a set union is:

set = *set* [+ ...]

For example, suppose sets SR (supply regions) and DR (demand regions) have been defined. Then, one may specify:

```
LO = SR + DR
```

to be the set of all regions.

The syntax for a numeric set is given by:

set = *d1*,...,*d2*

where *d1* and *d2* are either numeric constants or parameters. For example,

```
t = 1,...,HORIZON
```

defines the set t to be the integers {1,...,HORIZON}, where HORIZON is a parameter. The ellipsis distinguishes the numeric set.

Once sets are defined, *domains* are defined as cross products of sets, which may be restricted to delineate only certain members of the product.

For example, a table COST(SR,DR), is defined over the domain SR×DR, which can be the unit cost to transport something from a supply region to a demand region. An activity, P(SR,MT), can represent production of some material in the (declared) set MT at some supply region (SR). Perhaps only some supply regions produce a particular material, so the domain can be restricted to certain combinations of supply region and material. Syntactically, MODLER recognizes such restrictions as data-dependent conditions on the activity's generation by specifying:

`P(SR,MT):` ***`condition`***

where the condition is a list of relations, each of the form: ***`data_object relation data_object`***. Each data object can be a numeric constant, a parameter, or a table. Here is an example:

```
P(SR,MT): SUPPLY(SR,MT) > 0
```

In this case, the activity is generated for those supply regions and materials for which the associated entry in table SUPPLY is positive.

Domains can also be in the syntax of the definition of an equation. For example, we can specify:

```
S(SR,MT) = P(SR,MT) - T(SR,DR,MT)
```

The name of the equation is S (for supply), and its domain is SR×MT. As we examine the terms on the right side of the equal sign, we see the production activity, P(SR,MT), which is defined over the same domain, and the transportation activity, T(SR,DR,MT), whose domain has the projection SR×MT.

We shall be saying more about this in the next chapter, but for now notice that the equation's domain fixes associated set indexes to the right of the equal sign. That is, an equation will be generated for each combination of supply region (SR) and material (MT). For one particular combination, say (s,m), there is exactly one production activity in the equation, namely P(s,m), but there are many transportation activities: T(s,d,m) for all d in DR. The appearance of DR in the domain of activity T is not fixed by the equation's domain, so it is interpreted as *for all*. We also call the associated set index *free*.

In general, an equation has the syntax:

$$name[(s_1,\ldots,s_k)] = [-]\ term\ [\pm\ term\ [\ldots]],$$

where each term can have the simple form, *[data_object*]decision_object*, and $(s_1,\ldots,s_k)$, if present, is the domain of the equation (if absent, the domain of the equation is the empty set, such as a COST equation; at least one such equation is in the model, namely the objective). If the data object in a term is absent, it is assumed to be 1, as in the previous example, namely equation S(SR,DR).

In general, the data object and the decision object can have non-empty domains (in the former case, the data object is a table; otherwise, it is a parameter or a numeric constant). If a set appears in the domain of an equation, its associated index is considered fixed. All other sets on the right of the equal sign have associated set indexes that are free, which is equivalent to a summation in the equation. For example, suppose we have:

```
E(S1,S2) = T(S1,S3)*A(S1,S2,S3) + U(S4)*B(S4,S2) + C(S1) +
                      V(S1,S4)*D(S2,S5)
```

where S1, S2, S3, S4, and S5 are sets that have been defined; T, U, and V are tables that have been defined with domains as shown; and, A, B, C and D are activities that have been defined with domains as shown.

Algebraically, this is equivalent to:

$$E(i,j) = \Sigma_{k\in S3}\ T(i,k)A(i,j,k) + \Sigma_{k\in S4}\ U(k)B(k,j) + C(i) + \Sigma_{k\in S4,h\in S5}\ V(i,k)D(j,h)$$

for all $i\in S1$, $j\in S2$.

We shall illustrate this further with many examples, which came with your MODLER disk. For now, note that domains are defined as cross products of sets, and they comprise the cornerstone of the model's syntax.

2.3 Objects and relations.

The basic ingredients of any model are its objects and relations among its objects. We have already alluded to some of these in the previous section, and we shall illustrate in more detail in subsequent chapters. Here we formalize the definitions used by MODLER.

An object is either a data object or a decision object. A data object is either symbolic or numeric. The only symbolic data object in MODLER is a set, from which domains are defined. Because of their special role, sets are sometimes treated separately from the other data objects in discussing them; however, the members of sets are as much a data concept as numeric entries of a table. Numeric data objects are constants, parameters and tables. A constant can be integer or real-valued, with no distinction in MODLER. The only difference between a parameter and a table in MODLER is that a parameter has an empty domain – that is, it is a scalar, which does not depend on any set, whereas a table has at least one set in its domain.

Decision objects are either activities (with continuous-valued levels) or binary (whose levels are either 0 or 1). A decision object can have an empty or non-empty domain. In addition, its domain can be restricted by relations between data objects.

Relations in a model are of two types: conditions and constraints. Conditions include domain restrictions, as in defining decision objects, and admissible values for numeric data objects, as ranges of values for parameters and table entries. Constraints are either arithmetic or logical. Arithmetic constraints are equations, bounds, and limits. Logical constraints include implications (IF ... THEN ...) relations that involve binary decision objects.

Figure 2-2 gives a maero-view of these ingredients. Much more will be said about them in subsequent chapters.

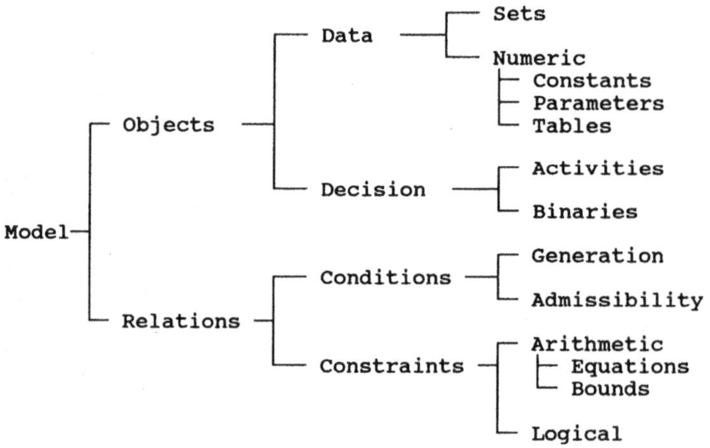

Figure 2-2. Anatomy of a Mathematical Programming Model

Chapter 3. Preliminaries

A computer does not substitute for judgement any more than a pencil substitutes for literacy. But writing without a pencil is no particular advantage.
Robert S. McNamara

In this chapter we start with the sign-on screen and describe some of the interactive conventions, such as abbreviations of commands and keyword parameters. Then, we shall illustrate how to query model objects and relations. The dialogue is designed for you to follow the exercises, and it is recommended you do so. Change to the directory where you installed MODLER. Then, at the DOS prompt enter:

> MODLER

3.1 The Sign-on and basic interactive conventions.

When you first enter MODLER, credits will appear, as shown in figure 3-1. This is simply printing the file CREDITS.DOC, which you may delete or rename to avoid printing in subsequent sessions. You answer the prompt simply by pressing Enter (Y is the default response).

```
                              CREDITS

MODLER was written by Dr. Harvey J. Greenberg, University of
Colorado at Denver.  The development was sponsored by a Consortium
of companies as part of a project to develop an "Intelligent
Mathematical Programming System."  The companies are:  Amoco Oil
Company, IBM, Shell Development Company Chesapeake Decision Sciences,
GAMS Development Corp., Ketron Management Science, MathPro Inc., and
Maximal Software.  Partial support was also provided by the Office of
Naval Research (contract #N-00014-88-K-0104).

Special help with coding and testing was provided by Milton M.
Gutterman (Amoco) and Thomas Dehne (Ketron).  The modulus and
initial value for the uniform random number generator in the FLIP
subroutine library was obtained from Dr. Richard E. Nance,
Virginia Polytechnic Institute.  Advice was given by Frederic H.
Murphy.

Dr. Greenberg is responsible for any errors in the programs but
assumes no liability, as this was written as part of an
"experimenter's workbench" to aid research in mathematical
programming.

...To see credits anytime, PRINT CREDITS.DOC
```

Figure 3-1. Initial Screen gives Credits

The Version shown in figure 3-2 may not agree with your version. This may mean you should obtain the latest, but the rest of this User's Guide should still be a valid introduction to MODLER. Note the message to PLEASE LOCK CAPS; MODLER is case-sensitive, and all commands and keywords require capital letters. The names of objects and relations in the model, however, may contain lower-case letters, and they must be entered as such.

```
                        MODLER  Version 6.1
                                  by
                         Harvey J. Greenberg

                    : Enter  ?  for introduction :
                    : Enter HELP to see commands :
                    :............................:


 PLEASE LOCK CAPS

                    ... Enter  ? ? for help with dictionary
MODLER ... H
        MODLER   has  16 session control commands:
    EXECUTE  HELP     OUTPUT   PRINT    QUIT     RETURN
    SCREEN   STRING   SWITCH   RECAP    _EJECT   _PAUSE
    _SETUP   _DICTNRY _KEYWORD _TYPE
        and  10 procedural commands:
    ATTRIBUT DISPLAY  EXPLAIN  LIST     RANDOMIZ READIN
    SCHEMA   SHOW     SUMMARY  WRITEOUT

   Enter HELP * to get help for each command
```

Figure 3-2. MODLER Sign-on and Command List (Response to HELP)

After the initial sign-on, you may enter ? for a general introduction; and, you may enter HELP (or simply H) to see the commands. If you do the latter, you will see the list of commands, as shown in figure 3-2, and you may enter any command after HELP for on-line help.

A Discrepancy

> On your PC the MODLER ... prompt appears on a line by itself, but figure 3-2 shows my response (H) on the same line. This appears throughout this User's Guide to save the extra line because some figures are rather long. Such edits have been minimized, and what you see in the User's Guide should agree with what you see on your screen. There are a few instances where some minor editing was done to accommodate the width of the User's Guide page, which is less than the width of your screen.

Among the conventions you will experience, MODLER (more accurately, the *F*ortran *L*anguage *I*nteractive *P*rocessor) allows telescopic abbreviations. The reason that H means HELP is that there is no other command beginning with H. Other telescopic abbreviations of HELP are HE and HEL. This carries to keywords and parameters within commands. For example, to read a model file into memory, the complete command specification is:

`MODLER ...READIN MODEL` *filespec*

We may, however, enter:

`MODLER ...REA M` *filespec*

The least abbreviation of READIN is REA because RETURN is another command that starts with RE. We may abbreviate the keyword (file type) MODEL as M, MO, MOD, or MODE. The *filespec* is a file name, but it may have a prefix (i.e., a path) or a suffix other than what has been established in the _SETUP command. For this introduction, we shall bypass file naming conventions and simply use the file name for any *filespec* we need. Later, in Chapter 8, we shall describe how to use the _SETUP command before and during a session to reset the default prefixes and suffixes of the various file types.

The commands are grouped functionally into two kinds: *session control* and *procedural*. The session control commands fulfill functions common to any interactive session, regardless of application; procedural commands are specific to MODLER functions. Most of the session control commands will enter our exercises naturally, so let's immediately begin to apply the procedural commands.

3.2 Elementary models and query.

To begin, consider the following linear program.

```
minimize COST:  P,T ≥ 0

COST     = 5*P + 2*T
S        =   P -   T = 0
D        =         T ≥ 25
```

Here the only data objects are the constants 1 (implicit coefficients), 5, 2, 25 and 0. The two decision objects, named P and T, may be regarded as Production and Transportation activities, respectively. The following specifies this model, with the constants "hard-wired."

```
NAME  MYPROB
* This line is a comment.
ACTIVITIES
  P
  T
EQUATIONS
  COST  = 5*P + 2*T
  S     =   P -   T
  D     =         T >= 25.
ENDATA
```

It is presumed that S=0 because no other specification appeared. Further, the activities P and T are presumed to be simply nonnegative. If we wanted to bound the level of production (P) by 100, we could declare this right where the activity is defined, or in the BOUNDS segment of the model file (we shall illustrate this below). It is also presumed COST is to be minimized. If, instead of COST, the first equation was named OBJECTIV (only 8 characters are allowed), then the LIMITS section could declare:

```
OBJECTIV  MINIMIZE
```

(or this could be just when the OBJECTIV equation is defined).

Suppose we want to run this model with different cost coefficients and with different demand values. Instead of making these data objects numeric constants, we can define parameters as follows.

```
NAME  MYMODEL
PARAMETERS
  CP
  CT
  DEMAND
ACTIVITIES
  P
  T
EQUATIONS
  COST = CP*P + CT*T
  S    =  P  -  T  = 0
  D    =          T >= DEMAND
ENDATA
```

(We showed S = 0 for documentation; it is already the default.)

Now we may want to define meanings of the objects for documentation and later analysis support. This is done by adding text after the declaration, as follows.

```
NAME  MYMOD
PARAMETERS
  CP cost of production
  CT cost of transportation
  DEMAND amount demanded
ACTIVITIES
  P produces
  T transports
EQUATIONS
  COST totals cost      = CP*P + CT*T  = MINIMIZE
  S    balances supply  =    P -    T  = 0
  D    requires demand  =           T >= DEMAND
ENDATA
```

(We showed COST = MINIMIZE for documentation; it is already the default.)

Values for the parameters are read from a data file. As part of model management, and for experimentation or prototyping, we may define ranges for the parameters. For example, we may declare:

```
CP cost of production: 1/100
CT cost of transportation: 0/*
DEMAND amount demanded: 10/1000
```

The parameter CP is required to be in the range 1 to 100 (inclusive); CT is required to be in the range 0 to infinity (*), or simply nonnegative; DEMAND is required to be in the range 10 to 1000. When data is read in, these values are checked, and the parameter value is rejected if it is not in its required range.

Ranges for data objects, such as parameters, may be a list. For example, suppose we define:

```
CT : 0, 10/*
```

This requires CT to be either 0 or in the range $[10,\infty)$. The general syntax for a range list specification is:

$$r_1 \ [,r_2 \ [\ldots]]$$

where each range (r_1, r_2, ...) is of the form,

```
v or v/u,
```

where v is a (numeric) value. If the slash (/) is used, v is the lower value of the range, and u is its upper value. Here are some examples of range specifications:

5/10	$[5,10]$
5	$= 5$
5/5	$= 5$
0/*	$[0,\infty)$
*	$= \infty$
0	$= 0$
-1/1	$[-1,1]$
-*/0	$(-\infty,0]$
-*/*	$(-\infty,\infty)$

where ∞ is an internal representation of infinity.

Now suppose the model is regional, where there are production and transportation activities for each of several supply regions (but still one demand location). Now the domain of these activities is defined by a set of regions. Figure 3-3 gives a model file for this representation.

```
NAME  MYMOD
PARAMETERS
  DEMAND amount demanded : 1/*
SETS
  SR supply region
TABLES
  CP(SR) cost of production in (SR) : 1/100
  CT(SR) cost of transportation from (SR) to demand location : 0/*
ACTIVITIES
  P(SR) produces in (SR)
  T(SR) transports from (SR) to demand location
EQUATIONS
  COST   totals cost        = CP(SR)*P(SR) + CT(SR)*T(SR) = MINIMIZE
  S(SR)  balances supply in (SR) =  P(SR) -        T(SR) = 0
  D      requires demand         =                 T(SR) >= DEMAND
ENDATA
```

Figure 3-3. A Model File for a Regional Production-Distribution Linear Program

The parameters CP and CT have become *tables* because they now depend on the set SR. Ranges of admissibility have been added to the model specification. The DEMAND parameter is required to be at least 1, indicated by the range specification, 1/*.

The table CP(SR) is defined over the domain composed of the set of supply regions (SR), and its range of admissibility is [1,100]. Its meaning has been expanded to include the reference to set SR with the phrase: `cost of production in (SR)`. Similarly, the table CT(SR) is defined over the same domain, and its range of admissibility is $[0,\infty)$. Its meaning has been similarly expanded.

The activities P(SR) and T(SR) now have domains SR, and we may regard P and T as *activity classes*. Their meanings have been expanded to reference their domains. This can be used not only for documentation, but also to pass this syntax on to the ANALYZE system, which gives English translations of the activities that are generated for each instance.

Now let us consider the equations. There are three equations shown, COST, S and D. The supply equation (S), however, is an *equation class*, defined over the domain composed of the set of supply regions (SR). Let us examine this and how MODLER uses the set references in the domains of the tables and activities to form the appropriate summations.

Since COST has no set references, the domains (SR) on the right side of the equal sign says that *every* production and transportation activity appears in the equation. This is equivalent to the sum:

$$COST = \Sigma_{i \in SR} CP(i)P(i) + \Sigma_{i \in SR} CT(i)T(i).$$

The same principle applies to the demand equation (D). Because there are no set references in the domain of D, every transportation activity appears in the equation. The net result is that the value of D is the sum of the transportation levels:

$$D = \Sigma_{i \in SR} T(i).$$

Equation S, however, has domain (SR). The specification says, for each supply region – that is, for each member of SR – there is an equation that balances production and (outbound) transportation for that region. There is therefore no summation; each supply equation has one production activity and one transportation activity.

To illustrate, figure 3-4 lists an instance of the model with the following data:

```
DEMAND = 100
SR = {NE, SW}
```

SR	CP	CT
NE	11	12
SW	21	22

```
COST = 11*PNE + 21*PSW + 12*TNE + 22*TSW = MIN
SNE  =    PNE          -    TNE          = 0
SSW  =           PSW              -  TSW = 0
D    =                      TNE +    TSW >= 100
```

Figure 3-4. An Instance of the Model in Figure 3-3

To recapitulate, a model is a set of objects plus relations among the objects. We have 2 types of objects: data objects and decision objects. Data objects include numeric values, which could appear literally as constants or symbolically (to be given a value later) as parameters and tables. The only difference between a parameter and a table is that the former has no set references. Decision objects are either activities (with continuous-valued levels) or binaries (0 or 1). Each decision object may or may not have set references.

We shall say more about relations in the next chapter. So far, the only relation in the model specification we have discussed is the equation, which has the form:

$$E[(s)] = [-]\ term_1\ [\pm\ term_2\ [\ldots]]$$

where E is the name of the equation, s is a tuple of its set references that comprise its domain (which may be absent), and $term_1$, $term_2$, ... are its terms. At least one term is required. A term can have the simple form:

*[data object]*decision object*

If the data object is omitted, it is presumed to be 1.

Another type of relation, which is part of the model's logic, is the range of admissibility of each numeric data object. This seems entirely different than an equation in so far as calling it a relation in the model. Yet another type of relation, which also seems different, is a conditional specification for generation of a decision object, such as an activity.

On-line help for the examples in this section is available with SYNTAX.DOC, MODEL.DOC and the HELP command. You can also PRINT LESSONS for a guide.

Let us proceed with an extension of what we have done so far by considering the *generalized transportation problem*. Algebraically, this is given as:

$$\text{Minimize}\quad COST = \Sigma_{ij}\ t_{ij}T_{ij}:\quad T \geq 0,$$

Supply:	$\Sigma_j\ T_{ij} \leq s_i$	for all $i=1,\ldots,m$
Demand:	$\Sigma_i\ g_{ij}T_{ij} \geq d_j$	for all $j=1,\ldots,n$
Capacities:	$T_{ij} \leq U_{ij}$	for all i,j,

where i indexes a set of supply regions and j indexes a set of demand regions. The data elements are:

m ...number of supply regions (range of i);
n.....number of demand regions (range of j);
t .....unit transportation costs over (i, j);
g.....gain factors over (i, j), which must be positive;
U ...capacities over (i, j), which must be non-negative;
s .....supplies over i;
d.....demands over j.

Advice from a student

> From this point, through subsequent chapters, we shall examine models and data from files that came with MODLER. You might find it useful (though not necessary) to have a printed copy of these files. They have qualifiers .MOD, .DAT, and .INC. In addition, after you executed ATEST, matrix and syntax files were created, whose qualifiers are .MAT and .SYN, respectively. Print these, if you wish, using the DOS print command, and keep them handy in alphabetical order. (Further information about these file types are postponed until chapter 8, but their meanings will become clear as we go through examples.)

Figure 3-5 presents a model file for the problem as stated. Unlike the previous set definition, the supply and demand regions are specified as sets of integers. This is indicated with the MODLER syntax:

```
set [meaning] =  a ,..., b
```

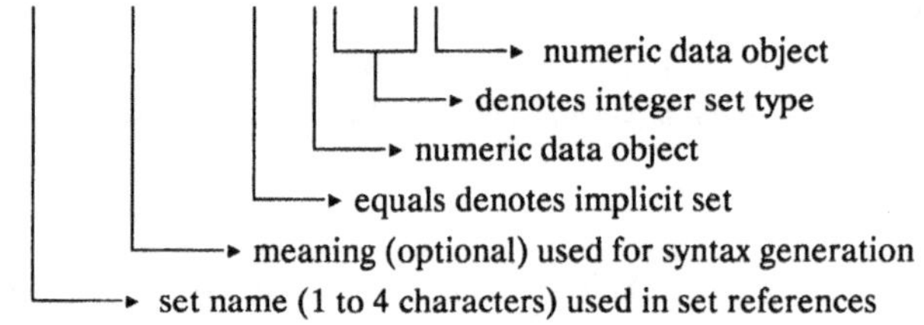

```
NAME     TRANSGEN
PARAMETERS
  M number of supply regions : 1/9
  N number of demand regions : 1/9
SETS
  i = 1,...,M
  j = 1,...,N
TABLES
  t(i,j) transportation costs
  g(i,j) gain from (i) to (j) : .001/*
  U(i,j) capacities : 0/*
  s(i)   supply at (i) : 0/*
  d(j)   demand at (j) : 0/*
ACTIVITIES
  T(i,j) transport from (i) to (j)
      <= U(i,j)
EQUATIONS
  COST      = t(i,j)*T(i,j)
  SUPPLY(i) =        T(i,j) <= s(i)
  DEMAND(j) = g(i,j)*T(i,j) >= d(j)
ENDATA
```

Figure 3-5. A Model File for the Generalized Transportation Problem

This version of the generalized transportation model illustrates some of the MODLER syntax. The data condition that the gain (g) be strictly positive is specified by a range [.001,∞), where .001 was chosen arbitrarily to be a small positive value that constitutes a least value for each gain factor. We also limited the number of supply and demand regions to be no greater than 9 in each case by the range specification 1/9 for their parameters, M and N, respectively. If more than 9 regions are possible in an instance, the set names (i and j, respectively) need to have two characters, like SR and DR.

In general, the length of a set name restricts the length of its members. If a set name has only one character, each member can have only one character. Members of integer sets are the integer symbols, 1, 2, ... The tenth member, therefore, would require two characters (10). The set could be named SR, say, in which case it can have up to 99 members. The same property applies to explicit sets, whose members are defined by a list in a data file.

To some extent we are getting a bit ahead of ourselves. If some confusion has crept in, temporarily ignore it. We shall go through some prepared examples to illustrate many of the MODLER capabilities and limitations. Complete descriptions of model and data files are given in chapter 8.

Figure 3-6 shows a different form of the generalized transportation problem, which came with your MODLER disk. You can read this into memory by specifying:

READ M TRANSGEN

where this abbreviates the full specification,

READIN MODEL TRANSGEN.MOD.

The first noticeable difference is that the gain factor is uniform - that is, it does not depend upon the supply region or the demand region. Because its domain has no set references, it is defined as a parameter. Further, its range of admissibility is [0.9,1.1], which corresponds to a 10% loss when GAIN = 0.9, and a 10% gain when GAIN = 1.1.

The second thing to notice is that we added a set of materials (MT), so the problem is multi-commodity. This added dimension appears in the domains of the tables. For example, the supplies are given in the table SUPPLY(MT,SR). A third thing to notice is that its range list allows a supply of some material at some supply region to be 0, * (infinity), or in the range 100/1000. The allowance of 0 says some materials may not be available at some supply regions. The allowance of * says that some supplier may have (virtually) unlimited supply of some material. Otherwise, if there is some supply and it is not infinite, it must be at least 100 and at most 1000 units.

```
NAME      TRANSGEN   General Transportation Model with uniform gain
PARAMETERS
 GAIN volumetric gain or loss : .9/1.1
SETS
 SR supply region
 DR demand region
 MT material
TABLES
 SUPPLY(MT,SR)       available supply of MT at SR  : *, 0, 100/1000
 DEMAND(MT,DR)       required demand of MT at DR  : 0/*
 CAPACITY(MT,SR,DR) capacity for transporting MT from SR to DR: 0,
                                                              10/*
 TRANCOST(MT,SR,DR)  unit cost to transport MT from SR to DR
 PRODCOST(MT,SR)     unit cost to produce MT at SR
 REVENUE(MT,DR)      unit revenue to consume MT at DR : 0, 10/100

ACTIVITIES
 P(MT,SR)     produces (MT) at (SR):  PRODCOST(MT,SR) <> *
 T(MT,SR,DR) transports (MT) from (SR) to (DR) :  DEMAND(MT,DR) > 0,
                  TRANCOST(MT,SR,DR) <> *, CAPACITY(MT,SR,DR) > 0,
                  SUPPLY(MT,SR) > 0
 C(MT,DR)     consumes (MT) at (DR)

EQUATIONS
 S(MT,SR) limits supply of (MT) at (SR) = P(MT,SR) - T(MT,SR,DR)
 D(MT,DR) requires demand of (MT) at (DR) =     GAIN*T(MT,SR,DR) -  C(MT,DR)
 COST      totals cost    =   TRANCOST(MT,SR,DR)*T(MT,SR,DR) +
                      PRODCOST(MT,SR)*P(MT,SR) - REVENUE(MT,DR)*C(MT,DR)
BOUNDS
 T(MT,SR,DR) <= CAPACITY(MT,SR,DR)
LIMITS
 S(MT,SR) >= - SUPPLY(MT,SR)
 D(MT,DR) >=   DEMAND(MT,DR)
ENDATA
```

Figure 3-6. The TRANSGEN.MOD Model File

Now consider the activity specifications. In particular, the domain of the Production activity class is the cross product of materials and supply regions (MT,SR). Not every member of this product, however, will be present in an instance of the model. Table PRODCOST controls this by having * for those product members that have no production. This may be regarded as assigning infinite production cost to the unwanted activities, but MODLER actually suppresses their generation (rather than enter them with an infinite cost).

The Transportation activity has domain (MT,SR,DR), but there are three conditions for the generation of an activity in an instance:

`DEMAND(MT,DR) > 0,`

.................there must be demand for the material at the demand region

`TRANCOST(MT,SR,DR) <> *,`

.................the transportation cost must be finite

`CAPACITY(MT,SR,DR) > 0,`

.................there must be capacity for the transportation

`SUPPLY(MT,SR) > 0`

.................there must be supply of the material at the supply region

The equations are natural extensions of what we had before, except that the supply and demand limits are defined in the LIMITS segment of the model file. This is also true for the capacity bound on the transportation activities. You may define limits with equations and bounds with activities, as before, or you may have them in separate segments.

As before, each equation defines a domain that holds fixed to the right of the equal sign. The other set references, not fixed by the equation's domain, are implied summations, as follows.

$$S(m,s) = P(m,s) - \Sigma_{j \in J(m,s)} T(m,s,j)$$

$$\text{for all } m,s \in (MT,SR), \text{ where } J(m,s) = \{j \in DR: T(m,s,j) \text{ is generated}\}.$$

$$D(m,d) = \Sigma_{i \in I(m,d)} GAIN*T(m,i,d) - C(m,d)$$

$$\text{for all } m,d \in (MT,DR), \text{ where } I(m,d) = \{s \in SR: T(m,s,j) \text{ is generated}\}.$$

$$COST = \Sigma_{(m,s,d) \in T} TRANCOST(m,s,d)*T(m,s,d) +$$

$$\Sigma_{(m,s) \in P} PRODCOST(m,s)*P(m,s) - \Sigma_{(m,d) \in C} REVENUE(m,d)*C(m,d),$$

$$\text{where } T = \{(m,s,d): T(m,s,d) \text{ is generated}\},$$

$$P = \{(m,s): P(m,s) \text{ is generated}\},$$

$$C = \{(m,d): C(m,d) \text{ is generated}\}.$$

In figure 3-7 we first read in the model file shown in figure 3-6. Then, LIST EQU is the command to list its equations. We see the three equation classes listed, as they were defined in the model file. Use the **EXPLAIN** command to see an explanation of this model.

```
MODLER ...REA M TRANSGEN
  (messages not shown)
MODLER ...LIST EQU
  S(MT,SR) =  P(MT,SR) - T(MT,SR,DR)  >=-SUPPLY(MT,SR)
  D(MT,DR) =  GAIN*T(MT,SR,DR) - C(MT,DR)  >=DEMAND(MT,DR)
  COST =  TRANCOST(MT,SR,DR)*T(MT,SR,DR) + PRODCOST(MT,SR)*P(MT,SR)
         - REVENUE(MT,DR)*C(MT,DR)  = MIN
MODLER ...DISPLAY PARAM //E
Parameter GAIN...volumetric gain or loss
         Admissible values:   0.9/1.1
         Current value: undefined
```

Figure 3-7. Reading and Querying the TRANSGEN.MOD Model File

Then, the parameter is displayed with the **DISPLAY PARAM** specification. The //E specification is the extended format; without it, the brief format (one line) is displayed (try it).

In figure 3-8 the activities are displayed. For each activity, its name, domain and meaning are given on the first line of its display. Then, the conditions for generation are given (following `When:`), followed by the activity's bounds.

```
MODLER ...D A //E
Activity P(MT,SR) ...produces (MT) at (SR)
         When:  PRODCOST(MT,SR) <> infinity
         Bounds:   >=  0 (no upper bound)
Activity T(MT,SR,DR) ...transports (MT) from (SR) to (DR)
         When:  DEMAND(MT,DR) > 0,
                TRANCOST(MT,SR,DR) <> infinity,
                CAPACITY(MT,SR,DR) >  0,
                SUPPLY(MT,SR) >  0
         Bounds:   >=  0  AND <=  CAPACITY(MT,SR,DR)
Activity C(MT,DR) ...consumes (MT) at (DR)
         When: always
         Bounds:   >=  0 (no upper bound)
```

Figure 3-8. Displaying the Activities in the Generalized Transportation Model

In figure 3-9, the model is listed (in algebraic form). In figure 3-10, the schema form is displayed.

```
MODLER ...LIST MODEL
                                   Model TRANSGEN
                    General Transportation Model with uniform gain
 Minimize COST
   Subject to:

  COST =  TRANCOST(MT,SR,DR)*T(MT,SR,DR) + PRODCOST(MT,SR)*P(MT,SR)
         - REVENUE(MT,DR)*C(MT,DR)
  S(MT,SR) =  P(MT,SR) - T(MT,SR,DR) >=-SUPPLY(MT,SR)
  D(MT,DR) =  GAIN*T(MT,SR,DR) - C(MT,DR) >= DEMAND(MT,DR)
   Decision variables:
                              0 <= P
                              0 <= T <= CAPACITY
                              0 <= C
```

Figure 3-9. Listing the Model

```
MODLER ...SCHEMA DISPLAY TABLE
            :::::::::::::::: BLOCK SCHEMA :::::::::::::::::
              P(MT,SR)      T(MT,SR,DR)  C(MT,DR)
  S(MT,SR)    1             - 1                       >=-SUPPLY
  D(MT,DR)                  GAIN         - 1          >=DEMAND
  COST        PRODCOST      TRANCOST     -REVENUE     ...MIN
  BOUNDS      0              0           0
              *             CAPACITY     *
```

Figure 3-10. Displaying the Model Schema Table

In figure 3-11 we read in an associated data file, TRANSGEN.DAT. The *filespec* in the READIN DATA is omitted, so the default name is used. (Your screen shows confirmation messages that are omitted in figure 3-11.) Then, the sets are listed. In the model file each of the three sets (SR, DR and MT) are explicit, but in the data file, set DR is declared equal to set SR. You might PRINT TRANSGEN.DAT to see the data specifications shown on your screen.

```
MODLER ...READIN DATA
  (messages not shown)
MODLER ...L S
 Set SR   Source: TRANSGEN.DAT
   NE Northeast
   SE Southeast
 Set DR   Source: TRANSGEN.DAT= SR
   NE Northeast
   SE Southeast
 Set MT   Source: TRANSGEN.DAT
   RM raw material
   IM intermediate product
   FP finished product
```

Figure 3-11. Reading a Data File

It is important to realize that setting DR=SR in the data file is different from defining DR=SR in the model file. In the former case, the equality is for an instance of the model; in the latter case, it is the model's logic and applies to all instances.

At this point, it is appropriate for you to go through all of the query commands, DISPLAY, LIST and SUMMARY. Use the HELP command to see their complete syntax. You might also apply these to other model files whose names you can obtain by PRINT DIRFILES (we shall do more of this in the next chapter).

3.3 Randomization by command.

In this section we illustrate the RANDOMIZ command, which fills in missing data in order to generate an instance of a model. The purpose in doing so is to test its generation and analyze some of its characteristics. We shall go further into the use of randomization in the next chapter.

> If you are not continuing from where we left off, enter READ P TRANSGEN to read in the generalized transportation model and an associated data file (saved in packed format when you executed ATEST).

In figure 3-12, we try to write a *matrix file* with the WRITEOUT command, and we are told that the gain parameter has no value, thus preventing this to be done. The subsequent DISPLAY confirms this.

```
MODLER ...WRITEOUT MATRIX
 ** DATA OBJECT GAIN     HAS NO VALUE
MODLER ...D P //E
Parameter GAIN...volumetric gain or loss
         Admissible values:   0.9/1.1
         Current value: undefined
```

Figure 3-12. An Unsuccessful Attempt to Write a Matrix File Due to an Undefined Parameter Value (GAIN)

All undefined data objects can be assigned values by the RANDOMIZ command. Figure 3-13 illustrates this. The value for GAIN is determined by a uniform distribution (the default) over its admissible values, which was specified by the condition .9/1.1 in the model file.

```
MODLER ...RAND
  (messages not shown)
MODLER ...WR M
   Writing matrix file for TRANSGEN
                  13        Rows
                  15        Columns
                  34        Nonzeroes in COLUMNS section
                  11        RHS values
                   0        Range values
                   4        Bound values
Matrix file written to TRANSGEN.MAT

WRITEOUT MAT complete
```

Figure 3-13. Using the RANDOMIZ Command to Assign a Value

We shall say much more about randomization in the next chapter. It is also possible to assign a value for GAIN from the terminal. This is illustrated in figure 3-14, where the syntax of the specification follows that of a data file (see §8.5).

```
MODLER ...REA D *
 ENTER DATA STATEMENTS FROM TERMINAL
 ...ENTER ENDATA WHEN FINISHED
PARAMETER GAIN = 1.05
 Reading parameter GAIN
ENDATA
 Data read from TERMINAL

MODLER ...D P //E
Parameter GAIN...volumetric gain or loss
        Admissible values:   0.9/1.1
        Current value:        1.050   from TERMINAL
```

Figure 3-14. Assigning a Value From the Terminal

Notice that the *source* of the current value is from TERMINAL. In general, the source of each data object is recorded and is part of the display in extended format. It gives the file, which could be TERMINAL, or it states it is from the RANDOMIZ command.

Chapter 4. Models and Instances

Discovering and defining problems, elaborating courses of action and making final choices are all stages in the decision-making process.

Herbert A. Simon

In this chapter we give more details on the use of MODLER with examples that came with the MODLER disk. After discussing some precepts, more examples are presented that illustrate most of the MODLER features for modeling linear programs. Then, we describe randomization to assign values to data objects.

4.1 Some precepts.

One of the philosophies underlying the design of MODLER is the separation of models and instances, which are defined by particular data values. The two ways of obtaining an instance of a model are by reading data, usually from a file, and by randomization. Both were introduced in chapter 3, and we shall consider them further in this chapter. (Details on data format options are given in §8.5.)

A model was defined in chapter 2 (cf., figure 2-2) by its anatomy. The first fundamental parts are objects and relations. These subdivide according to their properties and their roles in the model specification. The model file structure reflects this design philosophy, and a hierarchy of parts is defined as follows.

After the model name, the first object class is the *parameter*. This is a scalar, numeric data object. In chapter 3, the generalized transportation model was represented with a uniform gain as a parameter. The general form of a parameter specification is:

`name [meaning] [: range_list],`

where the ***name*** must be 1 to 8 characters without embedded blanks or delimiter symbols. If present, the ***meaning*** of the parameter is an aid to documentation; it can be up to 64 characters without a colon.

Recall that a ***range_list*** is a list of ranges. Each range has the form u[/v], where u and v are numeric constants or the special symbol * for infinity. If u and v are finite, the range is the interval [u, v]. The range u/* is the interval [u, ∞), and -*/v is ($-\infty$, v]. If /v is absent, it is assumed that v=u, so the interval is [u,u], which is just the point u, where u=* is allowed. If more than one range is present, they are separated by commas. For example, the range list 0, *, 10/1000 means that the parameter can be zero, infinity, or in the interval [10, 1000].

The range list is used for checking data specifications. If a parameter's value does not lie in one of the ranges in the list, it is rejected and an error message is given. The range list is also used for randomization, which we shall consider later. Overlapping ranges are valid, such as 0/100, 10/1000. Subsumed ranges are also valid, such as 0/100, 10/20.

A range list can be continued on the next line by specifying a comma as the last character on the line. This delimiter tells the parser to continue with range specifications.

The model file does not assign a value to a parameter unless the range list admits only one value. This is in keeping with the precept that the model file expresses the model, not a particular instance defined by data object value assignments.

After all parameters have been specified, sets are defined with the form:

```
name [meaning] [= set_expression],
```

where the *name* must be 1 to 4 characters without embedded blanks or delimiter symbols. If present, the *meaning* of the set is an aid to documentation and analysis support with ANALYZE; it can be up to 64 characters without an equal sign.

If the equal sign (=) is absent, the set is considered *explicit*, or *primitive*. Its members cannot be defined in the model file, as that is considered a property of an instance. A data file provides a list of members for an explicit set. Alternatively, the members can be specified by randomization.

If an equal sign is present, the set is considered *implicit*, or *derived*. It can depend on other sets already defined or on parameters. There are two types of *set_expression* that can be used. The first is a set union, which has the form:

```
set [ + set [+ ...]]
```

For example, suppose sets A and B have been defined. Then, we could have:

```
C = A
D = A + B
```

The second type of *set_expression* is a sequence of the form:

```
data_object,...,data_object,
```

where the *data_object* here is limited to numeric constants and parameters. The set is composed of the integers in the indicated range. For example, `S = 1,...,9` defines the set of nine integers. If HORIZON has been declared a parameter, the set t = 1,...,HORIZON is the set of integers from 1 to the value of HORIZON, which could vary instance to instance.

After all sets are defined, tables are declared. A table is a numeric data object with a domain consisting of at least one set. The explicit form of the declaration is:

```
name(domain) [meaning] [: range_list],
```

where the ***name*** must be 1 to 8 characters without embedded blanks or delimiter symbols. If present, the ***meaning*** is an aid to documentation; it can be up to 64 characters without a colon.

The ***range_list*** is the same for a table as for a parameter. It serves the same purpose of checking data and giving a numerical range for randomization. The ***domain*** is a cross product of sets expressed as an ordered list, (*set* [, *set* [...]]). For example, T(i, j) is a table whose name is T and whose domain is i×j.

Parameters, sets and tables comprise the data objects in a model file, and none are required. Next, decision objects are defined, and at least one is required for the model file to be valid.

The first decision object is an activity. This is a continuous-valued variable, or class of variables, defined by:

```
name[(domain)] [meaning] [: condition_list]
                [bound_spec],
```

where the ***name*** must be 1 to 8 characters without embedded blanks or delimiter symbols. If present, the ***meaning*** is an aid to documentation and analysis support with ANALYZE; it can be up to 64 characters without a colon or a relational token, like an equal sign.

If present, a ***condition_list*** is a conjunction of logical expressions that must be true for the activity to be generated in an instance. A condition has the form:

```
data_object relation data_object,
```

where each data object can be a numeric constant, parameter or table; and, the relation is one of the usual numeric relations: <, <=, =, <>, >=, >. For example, P > 0 is a valid condition if P is a parameter, which requires its value to be positive for the associated activity to be generated in the instance.

A list uses the comma to separate each condition, which is interpreted conjunctively. For example, consider the following activity specification.

```
T(SR,DR) : COST(SR,DR) <> *, SUPPLY(SR) > 0, DEMAND(DR) > 0
```

The activity class has name T and domain SR×DR, which are sets that were already defined. The first condition requires that the associated entry in the COST table be finite; the second requires that the associated entry in the SUPPLY table be positive; and, the third requires that the associated entry in the DEMAND table be positive. All three tables had to have been defined, and all conditions must hold true for a particular activity in the class. By this we mean that, for $s \in SR$ and $d \in DR$, Tsd is an activity if the conditions hold for this member, $(s, d) \in SR \times DR$.

If a table in a condition has a set in its domain that is not in the domain of the activity, this is interpreted as a universal quantification. For example, suppose we have:

```
T(SR,DR) : DEMAND(MT,DR) > 0
```

This means that for $(s, d) \in SR \times DR$, activity Tsd is generated if $\Sigma_{m \in MT}$ DEMAND(m, d) > 0.

The underlying precept, which also applies to equations (given below), is that of *binding* domain sets of one object to that of its *referent* object. In this case, the referent object is the activity T whose domain is $SR \times DR$. All references to SR and DR in a condition use *bound* set members - that is, they are bound to, or *fixed* by, the referent's domain. Any set in the domain of a table in a condition that is not fixed is considered *free*. A free set reference, say S, is interpreted as a summation.

Like a range list, a condition list can be continued on the next line by specifying a comma as the last character on the first line.

Bounds for an activity can also be defined at the same time as the activity is defined with the ***bound_spec*** option, but it must appear on the next line. This will be illustrated in the next section and formally defined in chapter 8.

The second decision object is a binary variable, which has the same form as an activity, except it does not have the ***bound_spec*** option. Its value in the mathematical program is required to be either 0 or 1, so no bound specification is valid.

After the data objects and decision objects are defined, the model's constraints are expressed in algebraic form. The first type of constraint is the equation, which has the form:

name[(***domain***)] [***meaning***] = [-] ***term*** [± ***term*** [...]] [***limit_spec***],

where the ***name*** must be 1 to 8 characters without embedded blanks or delimiter symbols. If present, the ***meaning*** of the table is an aid to documentation and analysis support with ANALYZE; it can be up to 64 characters without a relational token, like an equal sign.

Each term of the equation can be one of three forms:

```
Simple form:        [data_object*]decision_object
Reference form:     [data_object*]decision_object(domain spec)
Summation:          SUM[summand]
```

Wherever a data object is optional, its default value is 1. Our focus here is on the simple form. The reference form is illustrated in Chapter 5, and the summation is described in Chapter 6.

The limit specification can be either a bound specification or an optimization specification: = `MINIMIZE | MAXIMIZE`. The default specification is = 0 – that is, the equation is homogeneous in the constraint of the mathematical program. An exception to this default is when the equation has no domain and is one of the special names: COST, PROFIT, or REVENUE. The default limit specification for a COST equation is = MINIMIZE, and the other two have default = MAXIMIZE. We also defer elaborating on this until we go through some examples. All defaults are over-ridden by a limit specification. For example, if COST is a balance equation, not a minimand, then COST = ... = 0 defines the associated constraint (where ... means the other specifications, notably the terms, precede the limit specification, = 0).

There is one special point about equations and how summations are represented, which pertains to domain binding similar to those of conditions for activity generation. If a set S is part of the domain of an equation, it is considered *fixed* in each term to the right of the equal sign. If it appears in the domain of a table or decision object, there is no summation over that set. If a set is not fixed by the domain of the equation, it is called *free*. If a free index appears in a term, a summation is implied.

For example, consider the equation given by:

```
E(S1, S2) = T(S1, S3)*A(S3) + V(S2)*B(S2, S4),
```

where S1, S2, S3 and S4 are sets, T and V are tables, and A and B are activities. In mathematical notation, this is equivalent to:

$$E_{ij} = \Sigma_{k \in S3} T_{ik} A_k + \Sigma_{k \in S4} V_j B_{jk} \text{ for all } i \in S1 \text{ and } j \in S2.$$

A way to view this notion of fixed and free indexes is to think of activities *intersecting* equations. Then, if activity A(I, K) intersects equation E(I, J), this means for each (i, j)∈I×J (the domain of E), generate an equation, say Eij. The particular equation Eij is intersected by activities Aik *for all* k∈K. The universal quantification, *for all*, creates a summation over K.

Another view is that sets in the domains of the terms are ***bound*** to the domain of the equation, if present. This is explained with examples in the next section.

4.2 More examples.

We already examined the generalized transportation model and an instance of it in chapter 3. Let us continue with a few more examples.

In figure 4-1 we read in an assignment model. Then, we list its equations and display everything (indicated by *). Figure 4-2 shows the model file, itself, which you can see on-line by specifying **PRINT ASSIGN.MOD.**

```
MODLER ...REA MOD ASSIGN
 (messages not shown)
MODLER ...L E
   COST =   COST(pe,jo)*X(pe,jo) = MIN
   PER(pe) =   X(pe,jo) <= 1.0
   JOB(jo) =   X(pe,jo) >= 1.0
MODLER ...D *
SET pe person
SET jo job
TABLE COST(pe,jo)  cost to assign (pe) to (jo)
BINARY X(pe,jo)  assigns (pe) to (jo)
EQUATION COST
EQUATION PER(pe)  limits (pe) to perform 1 job
EQUATION JOB(jo)  requires (jo) to be performed
```

Figure 4-1. An Assignment Model

Notice the implied summations in the equations. For each person (pe), the domain of the equation, PER(pe) = X(pe, jo), fixes the member of set pe; however, jo is a free set, so we sum over all jobs. A way to view this is that activity X *intersects* equation PER *for all* jobs for the same member of set pe, common to the domains of X and PER. The *for all* interpretation is equivalent to a sum of terms, where the index of summation is over the set (jo) not fixed by the domain of the equation. Similarly, equation JOB(jo) leaves pe a free set, so X(pe, jo) is summed over pe.

```
NAME  ASSIGN Assignment of personnel to jobs
SETS
 pe person
 jo job
TABLES
 COST(pe,jo) cost to assign (pe) to (jo): 0/*
BINARIES
*    Note: the following generation condition suppresses unacceptable
*          assignments, defined by infinite cost
 X(pe,jo) assigns (pe) to (jo): COST(pe,jo) <> *

EQUATIONS
 COST    = COST(pe,jo)*X(pe,jo)
 PER(pe) limits (pe) to perform 1 job  = X(pe,jo) <= 1
 JOB(jo) requires (jo) to be performed = X(pe,jo) >= 1
ENDATA
```

Figure 4-2. The Model File for the Assignment Model

The model file, and the equation listing with the DISPLAY and LIST commands, can be regarded as an algebraic view of the model. Another view is a block schematic. Figure 4-3 shows how you can use the SCHEMA command to display a block schematic of the model. Each decision object, notably activity classes, is a column strip, and the last column strip is the constraint relations. Each constraint relation, notably equations, is a row strip, and the last two row strips are bounds.

```
MODLER  ...  SCH D
         :: BLOCK SCHEMA :::
           X(pe,jo)
  COST     COST      ...MIN
  PER(pe)   1        <= 1
  JOB(jo)   1        >= 1
  BOUNDS   0
           1
```

Figure 4-3. Using the SCHEMA Command to Display the Block Schema View

Now consider the product distribution model in figure 4-4. After reading in the model file and an associated data file, we list and display its equations (in extended format).

```
MODLER ...READ MOD PRODIST
  (messages not shown)
MODLER ...READ DAT
  (other messages not shown)
 ...INSTANTIATED TABLE _YIELDp
 ...INSTANTIATED TABLE _DEMANDm
 ...INSTANTIATED TABLE _DEMANDf
MODLER ...L E
   COST =   O&M_COST(p,PL)*P(p,PL) + SHIPCOST(PL,m,FP)*T(PL,m,FP) = MIN
   A(RM) =   CONSUME(RM,p,PL)*P(p,PL) <= SUPPLY(RM)
   C(PL) =   P(p,PL) <= CAPACITY(PL)
   B(PL,m,FP) =   YIELD(p,PL,FP)*P(p,PL) - T(PL,m,FP) = 0
   D(m,FP) =   T(PL,m,FP) >= DEMAND(m,FP)
MODLER ...D E
EQUATION COST
EQUATION A(RM)  limits consumption of (RM)
EQUATION C(PL)  limits total production at (PL)
EQUATION B(PL,m,FP)  balances flow of (FP) from (PL) to (m)
EQUATION D(m,FP)  demands (FP) at (m)
```

Figure 4-4. A Product Distribution Model

The corresponding algebraic form in mathematical notation is given as follows. First, define arrays O, S, C, s, c, Y and d to correspond, respectively, to the tables O&M_COST, SHIPCOST, CONSUME, SUPPLY, CAPACITY, YIELD and DEMAND; and, to avoid confusing indexes with sets, define sets P and M to be p and m, respectively. Then,

Minimize COST:

$$\begin{aligned}
\text{COST} &= \Sigma_{p\in P,\, l\in PL}\, O_{pl}P_{pl} + \Sigma_{l\in PL,\, m\in M,\, f\in FP}\, S_{lmf}T_{lmf}; && && \\
A_r &= \Sigma_{p\in P,\, l\in PL}\, C_{rpl}P_{pl} && \le s_r && \text{for all } r\in RM; \\
C_l &= \Sigma_{p\in P}\, P_{pl} && \le c_l && \text{for all } l\in PL; \\
B_{lmf} &= \Sigma_{p\in P}\, Y_{plf}P_{pl} - T_{lmf} && = 0 && \text{for all } l\in PL,\ m\in M,\ f\in FP; \\
D_{mf} &= \Sigma_{l\in PL}\, T_{lmf} && \ge d_{mf} && \text{for all } m\in M,\ f\in FP.
\end{aligned}$$

where the variables, $\{P_{pl}:\ p\in P,\ l\in PL\}$ and $\{T_{lmf}:\ l\in PL,\ m\in M,\ f\in FP : S_{lmf} \text{ is finite}\}$, are non-negative. (The generation condition for T is not shown in figure 4-3, nor are the bounds. These can be seen either by examining the model file or by displaying the activities in extended format.)

In figure 4-5, we display all of its data objects. The instantiation of tables _YIELDp, _DEMANDm and _DEMANDf, indicated in figure 4-4, occurs because these are *implied tables*. These will be defined fully in Chapter 7. Here they are defined in terms of other tables. For example, table _YIELDp is defined = YIELD(p,PL,FP), which is the sum of yields over processes (p). The purpose of these three special tables is to check the sums (see files PRODIST.MOD and PRODIST.CHK).

```
MODLER ...D P,S,T
  Parameter NUMPROC number of processes  =  3.00
  SET RM raw material
  SET  p  process = 1,...,NUMPROC
  SET  PL plant (location)
  SET  FP finished product
  SET  m market
  TABLE CONSUME(RM,p,PL)  amount of (RM) used by (p) at (PL)
  TABLE YIELD(p,PL,FP)  yield of (FP) from (p) at (PL)
  TABLE SUPPLY(RM)  supply of (RM)
  TABLE O&M_COST(p,PL)  operation & maintenance cost of (p) at (PL)
  TABLE CAPACITY(PL)  total capacity of (PL)
  TABLE SHIPCOST(PL,m,FP)  unit shipping cost of (FP) from (PL) to (m)
  TABLE DEMAND(m,FP)  demand for (FP) at (m)
  TABLE _YIELDp(PL,FP) = YIELD(p,PL,FP)
  TABLE _DEMANDm(FP) = DEMAND(m,FP)
  TABLE _DEMANDf(m) = DEMAND(m,FP)
```

Figure 4-5. Displaying Data Objects in the Product Distribution Model

If you examine the model file, you will find that all sets, except p, are explicit. The members listed in figure 4-6 are specified in the data file, PRODIST.DAT. The set p, whose members are processes, is an implicit set defined by:

```
p = 1,...,NUMPROC
```

where NUMPROC is a parameter. Notice in figure 4-6 that the source shows the set expression for this implicit set. The source of each explicit set is the name of the data file.

```
MODLER ...L S
 Set RM   Source: PRODIST.DAT
   SL steel
   WD wood
 Set p     Source: = 1,...,NUMPROC
   1  1st process
   2  2nd process
   3  3rd process
 Set PL   Source: PRODIST.DAT
   CH Chicago
   HO Houston
   DE Denver
 Set FP   Source: PRODIST.DAT
   CH chair
   TB table
 Set m     Source: PRODIST.DAT
   S  South
   N  North
   E  East
   W  West
```

Figure 4-6. Listing the Sets of an Instance of the Product Distribution Model

A value of 3 is assigned to NUMPROC in the data file, and this caused set p to become defined as shown. Notice the syntax created by MODLER. The word `process` is inherited from the set meaning; the modifiers, `1st`, `2nd`, and `3rd`, were added by MODLER.

Figure 4-7 shows the block schema view. The cell entries describe the coefficients in the LP matrix. For example, the coefficient of the production activity, P(p,PL), where p is in the set of processes and PL is in the set of plants, is in the table O&M_COST for the associated domain entry (p, PL). The constant 1 appears in row strip C(PL) and column strip P(p, PL). This says that the coefficient of P(p, PL) in row C(PL) is 1, where the plants (PL) match. As can be seen in the equation listing, this is an implied summation over processes (p).

```
MODLER  ...   SCH D
              ::::::::: BLOCK SCHEMA :::::::::
              P(p,PL)     T(PL,m,FP)
  COST        O&M_COST    SHIPCOST     ...MIN
  A(RM)       CONSUME                  <=SUPPLY
  C(PL)        1                       <=CAPACITY
  B(PL,m,FP)  YIELD        - 1          = 0
  D(m,FP)                    1         >=DEMAND
  BOUND         0            0
                *            *
```

Figure 4-7. Block Schema View of the Product Distribution Model

In general, a cell entry is a (signed) data object, where blank denotes 0 – that is, no entry. In general, The data object is either a table reference, parameter or numeric constant. The coefficient value is obtained from matching the domains of the row and columns strips, then summing over the unmatched column sets. For example, consider the submatrix defined by row strip B(PL, m, FP) and column strip P(p, PL). For each plant (PL), the columns P(p, PL) have a coefficient in B(PL, m, FP) for each (m, FP). The result is a summation across processes (p). The coefficient value is equal to the entry in the table YIELD. As shown in figure 4-5, the domain of YIELD is p×PL×FP – that is, the cross product of the three sets. The members are fixed by the row's domain, so YIELD(p, PL, FP) is the coefficient of column P(p, PL) in row B(PL, m, FP), which can also be seen with the equation listing.

The right-hand sides of the constraints have domains conformal to the row strip. For example, consider the constraint A(RM) < = SUPPLY. If SUPPLY were a parameter, each constraint has the same upper bound. In this model, however, SUPPLY is a table. As shown in figure 4-5, the domain of SUPPLY is the set RM (raw materials), so the constraint is actually $A(r) \leq SUPPLY(r)$ for each $r \in RM$. The reason the domain of SUPPLY is suppressed in the schema display is to save space in order to fit the row strip on one line.

Now consider our last example in this section, shown in figure 4-9. This is a product mix model, as indicated in the summary.

```
MODLER ...REA MOD PRODMIX
  (messages not shown)
MODLER ...REA DAT
  (messages not shown)
MODLER ...SUM
     Summary of model PRODMIX ...Make final products from raw materials
 Source file:  PRODMIX.MOD
   0 Parameters     3 Sets          5 Tables     1 Activities
   0 Binaries       3 Equations     0 Logical constraints
MODLER ...L E
  M(RM) =    MATLREQD(RM,FP)*P(FP)  <= MATERIAL(RM)
  L(P) =    PROCREQD(P,FP)*P(FP)  <= CAPACITY(P)
  PROFIT =    PROFIT(FP)*P(FP)   = MAX
```

Figure 4-9. A Product Mix Model

The sets and activities are displayed in figure 4-10 (in extended format). There is only one activity class, as was evident in the equations listed in figure 4-9.

```
MODLER  ...  D S,A //E
 Set RM has 3 members...raw material
     From PRODMIX.DAT
 Set P has 4 members...process
     From PRODMIX.DAT
 Set FP has 4 members...finished product
     From PRODMIX.DAT
 Activity P(FP) ...produce (FP)
          When: always
          Bounds:    >=  0 (no upper bound)
```

Figure 4-10. Displaying the Sets and Activities of the Product Mix Model

Figure 4-11 displays the tables. The extended format lets us see the admissible values and the source of the current values, which in this instance is the data file PRODMIX.DAT. In figure 4-12 we list one of the tables, MATLREQD, which is presented in rectangular format.

```
MODLER ...D T //E
 Table MATERIAL(RM) ...amount of (RM) available
         Admissible values:      0/*
         Current values from PRODMIX.DAT
 Table CAPACITY(P) ...amount of (P) capacity
         Admissible values:      0/*
         Current values from PRODMIX.DAT
 Table MATLREQD(RM,FP) ...amount of (RM) to make 1 unit of (FP)
         Admissible values:      0/*
         Current values from PRODMIX.DAT
 Table PROCREQD(P,FP) ...amount of (P) to make 1 unit of (FP)
         Admissible values:      0/*
         Current values from PRODMIX.DAT
 Table PROFIT(FP) ...unit profit from (FP)
         Admissible values:  any
         Current values from PRODMIX.DAT
```

Figure 4-11. Displaying the Tables of the Product Mix Model

```
MODLER ...L T MATLREQD
  Table MATLREQD(RM,FP)      Source: PRODMIX.DAT
                 MC            MT            PC            PT
      RO       15.000        10.000        16.000         9.000
      MA       14.000         8.000        15.000        10.000
      OK       13.000         7.000        14.000        12.000
```

Figure 4-12. Listing a Table for the Instance of the Product Mix Model

Figure 4-13 lists the sets (RM, FP) that comprise the domain of this table. This tells us what each of the member abbreviations means. For example, the member RO of set RM means rosewood.

```
MODLER ...L S RM,FP
 Set RM   Source: PRODMIX.DAT
   RO rosewood
   MA mahogany
   OK oak
 Set FP   Source: PRODMIX.DAT
   MC modern cabinet
   MT modern table
   PC provincial cabinet
   PT provincial table
```

Figure 4-13. Listing the Sets for the Instance of the Product Mix Model

4.3 Data objects.

The three basic data objects are: parameters, sets, and tables. A parameter is a numeric scalar, which comes before other data objects in the model file. Sets and tables can be *explicit* (sometimes called *primitive*) or *implicit* (sometimes called *derived*).

A simple explicit set is declared in the model file, and its members are entered by a list in the data file. Another explicit set is a *set map*. This maps a domain, which is a product of sets, to a range, which is a set. For example, suppose J is a set of journeys and C is a set of cities. Then, we could have the set map, ORIGIN(J) = C, which is a city for each journey, called the origin of the journey.

Set maps cannot be used in domain definitions of other objects, like tables, decision objects and constraints. They can, however, appear in conditions for sum expressions (see Chapter 6) and implicit tables.

An implicit set is defined in terms of parameters or other sets. It becomes instantiated once the objects that define it are instantiated, so it does not appear in a data file.

An explicit table is simply defined in the model file, and its values are entered in a data file (Chapter 8 describes the different formats that can be used). An implicit table is one that is defined by other data objects. Like an implicit set, an implicit table becomes instantiated once the objects that define it are instantiated, so it does not appear in a data file.

Implicit sets and tables are described fully in Chapter 7. On-line help is available with MAPS.DOC, SETS.DOC and TABLES.DOC.

4.4 Randomization options.

In order for the random values generated in your session to agree with those displayed here, you must enter MODLER anew. This is because we applied the RANDOMIZ command in chapter 3, which affects the sequence of random numbers henceforth generated.

Let us begin with the RANDOMIZ command. This is a quick way to assign values to all data objects not already assigned from a data file. Figure 4-14 begins by reading in the assignment model, then specifying the RANDOMIZ command.

```
MODLER ...READ M ASSIGN
  (messages not shown)
MODLER ...RAND
 Filled SET pe =          2
 Filled SET jo =          3
 Filled TABLE COST
 RANDOMIZATION COMPLETE
```

Figure 4-14. Reading the Assignment Model and Randomizing Data Object Values

Possible Discrepancy
The particular random values could differ from your session.

In figure 4-15, we list the sets. Notice that the meaning given to each member is RANDOMLY GENERATED, and the source is the RANDOMIZ COMMAND.

```
MODLER ...L S
 Set pe     Source: RANDOMIZ COMMAND
   01 RANDOMLY GENERATED
   02 RANDOMLY GENERATED
 Set jo     Source: RANDOMIZ COMMAND
   01 RANDOMLY GENERATED
   02 RANDOMLY GENERATED
   03 RANDOMLY GENERATED
```

Figure 4-15. Listing the Sets of Persons (pe) and Jobs (jo) Randomly Generated

Figure 4-16 lists the COST table whose entries were randomly generated from the uniform distribution over a truncation of its admissible range, [0, ∞), which was declared in the model file. The truncation of this infinite range, which can be controlled by an option of the RANDOMIZ command, defaults to [0, 1000].

```
MODLER  ...  L T
 Table COST(pe,jo)      Source: RANDOMIZ COMMAND
               01           02           03
     01      585.083      360.290      731.994
     02      659.423      941.711      453.552
```

Figure 4-16. Listing Table COST Randomly Generated from the Default Truncation of Its Admissible Range

The RANDOMIZ command uses defaults for everything except the range of values. More control over randomization is enabled by specifying options in a data file, which could be the terminal. Before explaining the latter, consider the syntax of the RANDOMIZ command:

RANDOMIZ [RANGE=*value*] [: *range*]

The RANGE value applies to each unassigned parameter and table value whose admissible range is infinite. The default is 1000. For example, if the admissible range is 0/*, the truncated range is 0/v, where v is the RANGE value specified in the RANDOMIZ command (or the default value of 1000 if not specified).

The *range* following the colon applies to explicit sets. If the range u/v is specified, the number of members assigned to each empty, explicit set is uniformly distributed over the integers, {u, u+1, ..., v}. The default range is 2/5. If /v is not specified, the number of members in the set is fixed at u. Implicit sets are instantiated once the parameters or sets defining them are instantiated.

Let us consider one final example of using the RANDOMIZ command. Figure 4-17 begins by reading in a model file, TABLEAU.MOD. As shown by the equation listing and display, this is a completely abstract algebraic model, as found in mathematical texts on linear programming, with no particular meaning to the parts.

```
MODLER ...REA MOD TABLEAU
  (messages not shown)
MODLER ...L E
   Z =   c(j)*X(j) = MAX
   Y(i) =   A(i,j)*X(j) >= b(i)
MODLER ...D E
 EQUATION  Z
 EQUATION Y(i)  row level for (i)
```

Figure 4-17. A Tableau Model

In figure 4-18, we display the parameters and sets. The parameter M is merely the number of rows, and N is the number of columns, in the abstract linear program. The sets of rows and columns are implicit, using the integers {1,...,M} and {1,...,N}, respectively.

```
MODLER ...D P,S //E
 Parameter M...number of rows
         Admissible values:   1/9
         Current value: undefined
 Parameter N...number of columns
         Admissible values:   1/9
         Current value: undefined
 Set i has 0 members...row index
    = 1,...,M
 Set j has 0 members...column index
    = 1,...,N
```

Figure 4-18. Displaying the Parameters and Sets of the Tableau Model

The tables, displayed in figure 4-19, are the coefficient matrix (A), the objective coefficients (c), and the right-hand side (b). There are no restrictions on their admissible values – that is, the model file did not declare any range lists for these tables.

```
MODLER ...D T //E
 Table A(i,j) ...coefficient matrix
          Admissible values:  any
          Current values undefined
 Table c(j) ...objective coefficients
          Admissible values:  any
          Current values undefined
 Table b(i) ...right-hand sides
          Admissible values:  any
          Current values undefined
```

Figure 4-19. The Tables of the Tableau Model

In figure 4-20, we apply the RANDOMIZ command. Since the range of parameters M and N are in [1, 9], the numbers of rows and columns are in this range. After a uniformly distributed realization is determined, the value is rounded to a nearest integer between 1 and 9, thus skewing the distribution away from the extremes. A realization rounds uniformly when it is in the interval, (1.5, 8.5); however, the probability of 1 member or 9 members is half of the other probabilities due to adherence to the original range. The realization shown in figure 4-20 has 2 rows and 6 columns (as before, the random values you see may not agree with those shown).

```
MODLER ...RAND
  (messages not shown)
MODLER ...L S
 Set i     Source: = 1,...,M
   1  1st row index
   2  2nd row index
 Set j     Source: = 1,...,N
   1  1st column index
   2  2nd column index
   3  3rd column index
   4  4th column index
   5  5th column index
   6  6th column index
```

Figure 4-20. Listing the Randomly Generated Sets of Rows and Columns in the Tableau Model

Figure 4-21 lists the tables that were randomly generated from the interval, [-500, 500]. This range was determined by the default RANGE value of 1000 and the original range of (−∞, ∞) for the table entries.

```
MODLER ...L T
  Table A(i,j)       Source: RANDOMIZ COMMAND
          i  j    Value
          -------------------
          1  1   -493.774
          1  2   -163.269
          1  3   -196.594
          1  4    126.464
          1  5    234.924
          1  6    200.744
          2  1   -197.021
          2  2    491.394
          2  3    237.609
          2  4   -21.85
          2  5    480.163
          2  6    466.247
  Table c(j)       Source: RANDOMIZ COMMAND
          j    Value
          ---------------
          1    12.329
          2    118.225
          3    69.274
          4    217.529
          5   -253.601
          6    117.004
  Table b(i)       Source: RANDOMIZ COMMAND
          i    Value
          ---------------
          1    67.382
          2    479.431
```

Figure 4-21. Listing the Randomly Generated Table Entries in the Tableau Model

Now let us consider randomization from a data file, which gives more options to control the values generated. In figure 4-22 we read in the packed file, created during the execution of ATEST, from the generalized transportation model file (TRANSGEN.MOD) and an associated data file (TRANSGEN.DAT).

```
MODLER ...READ P TRANSGEN
 READIN complete
```

Figure 4-22. Reading the (Packed) TRANSGEN Model Instance
(Created by ATEST from TRANSGEN.MOD and TRANSGEN.DAT)

A portion of the data file is shown in figure 4-23 (we shall consider data file formats in chapter 8). Three of the tables were randomized with the specification:

TABLE *name* RANDOMIZ [*options*]

where *name* is the name of the table, and the *options* are as follows.

ASSIGN *weight*...............Assigns probabilities for range selection
RANGE *value*.................Truncates random value
FRACTION *digits*...........Rounds value to limit fractional part
DISTRIB {*type*}.............Distribution for value selection

```
* Table CAPACITY is randomly filled with a prevalence of nonzero
* entries (90%).
TABLE CAPACITY RANDOMIZE  ASSIGN 1,9  FRACTION=0

* Table PRODCOST is randomized with a default range
* (note no range spec in model file).
TABLE PRODCOST RANDOM

* Table REVENUE is randomly filled (no 0's) using triangular
* distribution in its positive range:  10/100.
TABLE REVENUE RANDOM ASSIGN 0,1 DISTRIB TRIANG
```

Figure 4-23. A Portion of the Data File, TRANSGEN.DAT

To illustrate, consider the transportation data. Table CAPACITY is assigned entry values by the RANDOMIZ specification in the data file, as shown in figure 4-24. The ASSIGN option assigns weights 1 and 9 to the two range specifications that are in the model file: 0, 10/*, respectively. This means 0 is chosen with probability 10%, and the range 10/* is chosen with probability 90%. In the latter case, the interval $[0, \infty)$ is truncated to the RANGE default, [0, 1000]. As shown in figure 4-24, one of the 12 table entry realizations is zero; the others were taken from a uniform distribution on [0, 1000].

```
MODLER ...L T CAPACITY
  Table CAPACITY(MT,SR,DR)          Source: TRANSGEN.DAT
         MT  SR  DR   Value
         ----------------------
         RM  NE  NE   600
         RM  NE  SE     0
         RM  SE  NE   313
         RM  SE  SE   744
         IM  NE  NE   312
         IM  NE  SE   747
         IM  SE  NE   990
         IM  SE  SE   522
         FP  NE  NE   579
         FP  NE  SE   256
         FP  SE  NE   577
         FP  SE  SE   136
```

Figure 4-24. Listing the CAPACITY Table for the Realization of the Generalized Transportation Problem Data

The table is listed in relational format because its domain is the product of three sets, MT×SR×DR. The FRAC=0 option specifies that the values are to be integer. In general, the FRAC option of the RANDOMIZ specification in a data file is the number of digits in the fractional portion of the value. This is done by rounding.

The second table with the RANDOMIZ specification is PRODCOST, which uses default options. As shown in figure 4-25, unlike the CAPACITY table, table PRODCOST has no range list – that is, its admissible values are any.

```
MODLER ...D T //E
 Table SUPPLY(MT,SR) ...available supply of MT at SR
          Admissible values:      *,      0,  100/1000
          Current values from TRANSGEN.DAT
 Table DEMAND(MT,DR) ...required demand of MT at DR
          Admissible values:      0/*
          Current values from TRANSGEN.DAT
 Table CAPACITY(MT,SR,DR) ...capacity for transporting MT from SR to DR
          Admissible values:      0,  10/*
          Current values from TRANSGEN.DAT
 Table TRANCOST(MT,SR,DR) ...unit cost to transport MT from SR to DR
          Admissible values:  any
          Current values from TRANSGEN.DAT
 Table PRODCOST(MT,SR) ...unit cost to produce MT at SR
          Admissible values:  any
          Current values from TRANSGEN.DAT
 Table REVENUE(MT,DR) ...unit revenue to consume MT at DR
          Admissible values:      0,  10/100
          Current values from TRANSGEN.DAT
```

Figure 4-25. Displaying the Tables (in Extended Format) to See Their Admissible Range

Figure 4-26 lists table PRODCOST, which is presented in rectangular format. The entries are in the interval, [-500, 500], which reflects the uniform (default) distribution over that range.

```
MODLER ...L T PRODCOST
 Table PRODCOST(MT,SR)       Source: TRANSGEN.DAT
               NE             SE
     RM     580.993        611.206
     IM        *              *
     FP        *           217.834
```

Figure 4-26. Listing Table PRODCOST of the Generalized Transportation Model Instance

On the other hand, the REVENUE table has this specification in the data file (cf., figure 4-23):

```
TABLE REVENUE RANDOM ASSIGN 0,1 DISTRIB TRIANG
```

Now the ASSIGN option puts all of the weight in the second of its ranges: 10/100 (cf., figure 4-25). Further, instead of the uniform distribution over [10, 100], the triangular distribution is specified with the DISTRIB option. These are reflected in the realization listed in figure 4-27.

```
MODLER ...L T REVENUE
 Table REVENUE(MT,DR)       Source: TRANSGEN.DAT
                NE            SE
      RM       36.953       44.222
      IM       71.176       73.064
      FP       26.668       66.209
```

Figure 4-27. Listing Table REVENUE in the Generalized Transportation Model Instance

Further examples and details about the RANDOMIZ options are given in RANDOMIZ.DOC, which is available on-line. The default values are also given as:

ASSIGN *weight*...............Equal weights for each range in list
RANGE *value*..................1000
FRACTION *digits*........... 3
DISTRIB {*type*}.............Uniform

Chapter 5. Special Structures

Everything should be made as simple as possible, but not simpler.
Albert Einstein

This chapter describes how to use MODLER for some special structures that arise in many models within the scope of linear programming. The first extension shows how to represent binary variables and logical constraints. The second is an inventory model, where control tables are introduced as a way to express dynamic equations. The third case is approximation of nonlinear functions, notably a step-function approximation to supply and demand functions. Finally, an elastic programming representation of LP models is described.

5.1 Binary variables and logical constraints.

Binary variables are decision objects defined the same way as activities, except their value in the mathematical program is required to be 0 or 1. Generally, binary variables are used to represent propositions, such as the following examples.

Let x_j = 1 if project j is selected; else, 0.
Let x_{ij} = 1 if plant i is operational at time j; else, 0.
Let x_{ij} = 1 if job i is performed at time j; else, 0.

The binary variables appear in both equations and *logical constraints*, such as the following examples.

Select at most one project: $\Sigma_j x_j \le 1$.
Operate activities at plant i only if plant j is operational: $O_{ijk} \le U_{ijk} x_{ij}$ $\forall k$
(where U_{ijk} is the upper bound of activity O_{ijk}).
Job i must precede job k: $\Sigma_{t \le j} x_{it} \ge x_{kj}$ $\forall j$.

Such constraints could be defined as equations with associated limits, but MODLER allows direct definition of logical constraints with the form:

```
IF antecedent THEN consequent
```

where the antecedent must be a binary condition: *binary* = 0 | 1. The consequent can be either a binary condition or an activity bound condition: *activity* = L | U. Any activity that is in a logical constraint must be totally bounded (i.e., L and U must both be finite).

To illustrate, figure 5-1 reads an extension of the generalized transportation model, called TRANSLGL.MOD. As shown with the display, this contains a class of binary variables to represent opening a link between supply and demand locations.

```
MODLER ...READ M TRANSLGL
  (messages not shown)
MODLER ...D B
 BINARY O(SR,DR)  opens new link from (SR) to (DR)
MODLER ...L L
     L(SR,DR) : IF O(SR,DR)  =  0 THEN T(MT,SR,DR)  = L
```

Figure 5-1. A Generalized Transportation Model with Binary Variables and Logical Constraints

As with the case of equations, the domain of the logical constraint class fixes associated set indexes that appear in the IF ... THEN statement. In this case, the domain of the logical constraint is SR×DR, so there is one constraint for every combination of supply region (SR) and demand region (DR). For (s, d)∈SR×DR, the constraint is:

`IF O(s,d) = 0 THEN T(m,s,d) = L` *for all* m∈MT.

Since the lower bounds for T are all 0, MODLER generates the following inequalities (shown in mathematical notation) to represent this system of constraints.

$$\Sigma_{m \in MT} T_{msd} \leq O_{sd} \Sigma_{m \in MT} U_{msd} \text{ for all } s \in SR,\ d \in DR,$$

where $T_{msd} \leq U_{msd}$ are the upper bounds on T. Then, $O_{sd} = 0$ constrains $T_{msd} = 0$ for all m∈MT; and, $O_{sd} = 1$ results in the redundant constraint on the aggregate upper bound for transportation from s to d.

This constraint is reformed in the linear program as:

$$\epsilon \Sigma_{m \in MT} T_{msd} \leq O_{sd} \text{ for all } s \in SR,\ d \in DR,$$

where $\epsilon \leq 1/\Sigma_{m \in MT} U_{msd}$. More generally, Table 5-1 shows the linear inequality generated from a logical constraint (see LOGICAL.DOC, which is available with on-line help).

In Table 5-1, let z, z′ denote levels of binary decision objects, and let x denote the level of an activity.

Antecedent	Consequent	Constraint generated
$z = 0$	$z' = 0$	$-z + z' \leq 0$
	$z' = 1$	$-z - z' \leq -1$
	$x = L$	$-z + \epsilon x \leq \epsilon L$
	$x = U$	$-z - \epsilon x \leq -\epsilon U$
$z = 1$	$z' = 0$	$z + z' \leq 1$
	$z' = 1$	$z - z' \leq 0$
	$x = L$	$z + \epsilon x \leq \epsilon L + 1$
	$x = U$	$z - \epsilon x \leq -\epsilon U + 1$

where ϵ is a positive value such that $\epsilon(U\text{-}L) \leq 1$.

Table 5-1. Inequality Generated for Logical Constraint

Thus, we set $\epsilon = \min_j[1/(U_j - L_j)]$, where the min is performed over activities (j) that appear in a logical constraint. This imposes a restriction that activities in logical constraints must have both bounds explicitly present, and that $(U_j - L_j)$ cannot be too great – that is, we require $U_j - L_j \leq 1/\tau$, where τ is a small positive quantity, called a tolerance. We call such activities *totally bounded.*

Before we consider more examples to illustrate the entries in Table 5-1, consider the appearance of a logical constraint in a block schema.

Figure 5-2 shows the schema view, using the SCHEMA command. In this case, there is a row strip for the logical constraint, following the equations. One cell contains the antecedent, with the IF statement, and another contains the consequent, with the THEN statement.

```
MODLER  ...  SCH DISP TABLE
        :::::::::::::::::::::::: BLOCK SCHEMA ::::::::::::::::::::
          P(MT,SR)    T(MT,SR,DR)  O(SR,DR)
COST      PRODCOST    TRANCOST     LINKCOST  ...MIN
S(MT,SR)   1          - 1                       = 0
D(MT,DR)               1                      >=DEMAND
T(SR,DR)               1                      >=CONTRACT  & <=CAPACITY
L(SR,DR)              THEN = L     IF = 0
BOUND      0          TARGET+-     0
          SUPPLY      RANGE        1
```

Figure 5-2. Schema (Table) View with Logical Constraints

Notice, too, that the BOUND strips for the transportation activities, T(MT, SR, DR), represent the bound specification:

```
TARGET(MT,SR,DR) +- RANGE(MT,SR,DR)
```

The domain of each table, TARGET and RANGE, conforms with the domain of the activity class, so the bounds have the algebraic equivalent:

$$\text{TARGET}_{msd} - \text{RANGE}_{msd} \le T_{msd} \le \text{TARGET}_{msd} + \text{RANGE}_{msd}$$

for all m,s,d $\in$ MT×SR×DR (see model file TRANSLGL.MOD for the details).

Now let us illucidate the entries in Table 5-1. Suppose we have a single antecedent and a single consequent, which are both binary variables, say z and z′. Then, the linear inequality generated for each type of logical constraint is given by:

```
IF z=0 THEN z'=0  <==>  z' ≤ z      <==>  -z + z' ≤  0;
IF z=0 THEN z'=1  <==>  1-z' ≤ z    <==>  -z - z' ≤ -1;
IF z=1 THEN z'=0  <==>  z' ≤ 1-z    <==>   z + z' ≤  1;
IF z=1 THEN z'=1  <==>  1-z' ≤ 1-z  <==>   z - z' ≤  0.
```

The first equivalence says that $z' \le z$ forces $z'=0$ when $z=0$; and, it is no constraint when $z=1$. The second equivalence says that $1-z' \le z$ forces $1-z'=0$, which is the same as $z'=1$, when $z=0$; and, it is no constraint when $z=1$. The last two cases follow similarly. The point is that simple logical statements correspond naturally to linear inequalities with binary variables.

Suppose, however, z and z′ are not simply one binary variable, but an implied collection. For example, suppose we have the logical constraint, L(I,J):

```
IF A(I,K)=0 THEN B(J,H)=0,
```

where A is a binary decision object with domain I×K, and B is a binary decision object with domain J×H (I, J, K, and H are sets). For each (i, j)∈I×J, the logical constraint is of the form:

$$\text{IF } \forall k \in K.A(i,k)=0 \text{ THEN } \forall h \in H.B(j,h)=0.$$

In this case, $z = \Sigma_{k\in K} A(i,k)$ and $z' = \Sigma_{h\in H} B(j,h)$, so the generated inequality is:

$$\Sigma_{h\in H} B(j,h) \le \Sigma_{k\in K} A(i,k)\,|H|,$$

where $|H|$ = cardinality of H. When the antecedent is true – that is, $A(i,k)=0$ for all $k\in K$, the constraint forces $B(j,h)=0$ for all $h\in H$. When the antecedent is false, the right side of the inequality is at least $|H|$, which is a redundant upper bound on the sum of the binary levels on the left side.

Now consider the case where the consequent is of the form x=L – that is, the activity level (x) is forced to its lower bound. Suppose we have:

IF z=0 THEN x=L,

where z is a simple binary level and x is a simple activity level (no free sets).

From Table 5-1, the equivalent inequality is:

$$-z + \epsilon x \leq \epsilon L,$$

where L = lower bound of activity (perhaps 0), and ϵ is chosen to satisfy $0 < \epsilon \leq 1/(U\text{-}L)$ for U = upper bound of activity. In this case, when z=0, the constraint requires $x \leq L$. Since $x \geq L$, this forces x=L. When z=1, the constraint requires $x \leq L + 1/\epsilon$; however, since $\epsilon \leq 1/(U\text{-}L)$, this constraint is satisfied by the stronger inequality, $x \leq L + (U-L) = U$, which is redundant.

The extension to where the decision objects have free sets is similar to the previous case, where *for all* is the logical meaning of free set specification. Consider the following example, where the logical constraint is L(I,J).

IF A(I,K)=0 THEN X(J,H)=L.

Its logical meaning for $(i, j) \in I \times J$ is:

IF $\forall k \in K$.A(i,k)=0 THEN $\forall h \in H$.X(j,h)=L_{jh},

where L_{jh} is the lower bound value for X(j, h).

An equivalent inequality is:

$$\Sigma_{h\in H}\ (X(j,h)-L_{jh}) \leq\ (\Sigma_{k\in K}\ A(i,k))\ \Sigma_{h\in H}\ (U_{jh}-L_{jh}).$$

If A(i, k)=0 for all $k \in K$, the constraint, together with $X \geq L$, forces X=L. Otherwise, the right side of the inequality is at least equal to $\Sigma_{h\in H}\ (U_{jh}-L_{jh})$, which is redundant in the presence of $X \leq U$.

The inequality is rearranged, however, so that the coefficients of the antecedent binary variables are unity. In particular, the inequality generated is:

$$\epsilon\Sigma_{h\in H}\ (X(j,\ h)-L_{jh}) \leq\ \Sigma_{k\in K}\ A(i,\ k),$$

or

$$-\Sigma_{k\in K}\ A(i,k)\ +\ \epsilon\Sigma_{h\in H}\ X(j,h)\ \leq\ \epsilon\Sigma_{h\in H}\ L_{jh},$$

where $0 < \epsilon \leq 1/(\Sigma_{h\in H}\ (U_{jh}-L_{jh}))$. Thus, the coefficient of each binary variable is -1; the coefficient of each activity is ϵ; and, the right-hand side of the inequality constraint is $\epsilon\Sigma_{h\in H}\ L_{jh}$ (MODLER adds a small positive quantity to the right-side in order to prevent a false indication of infeasibility due to rounding errors).

If the antecedent is `IF A(I,K)=1`, the inequality changes by replacing A with its complement, 1–A, which is equivalent to:

$$\Sigma_{k\epsilon K}\ A(i,k)\ +\ \epsilon\Sigma_{h\epsilon H}\ X(j,h)\ \le\ \epsilon\Sigma_{h\epsilon H}\ L_{jh}\ +\ |K|.$$

If A(i,k) = 1 for all kϵK, X is forced equal to L. Otherwise, the inequality is redundant because it is satisfied by X=U.

Further information about how MODLER generates linear inequalities for logical constraints is given in LOGICAL.DOC, which is available on-line.

5.2 Dynamic models.

Consider an inventory equation of the form,

$$I_t = I_{t-1} + P_t - S_t,$$

where I_t = inventory at the end of period t; P_t=level of production at the beginning of period t; and, S_t=sales after production. One way this can be expressed in MODLER, using implied sums, is with the use of *control tables*.

A control table is one with binary values, 0 or 1, such that terms and conditions can be controlled with absence or presence, respectively, of a decision object. For example, suppose we declare in the model file:

```
PARAMETERS
  HORIZON planning horizon : 1/9
SETS
  T time period = 0,...,HORIZON
  t time period = 1,...,HORIZON
TABLE
  LAG1(T,t) control table where t is T-1 : 0, 1
```

In the data file we would set the table `LAG1` as:

```
TABLE  LAG1(T,t) = 0 EXCEPT
* T t Value
  0 1   1
  1 2   1
  2 3   1
  ...
```

where ... means this continues until t is the horizon. Figure 5-3 begins to illustrate this by reading in a model file called INVENTRY.MOD.

```
MODLER ...REA MOD INVENTRY
  (messages not shown)
MODLER ...SUM
                  Summary of model INVENTRY ...Inventory problem
  Source file:  INVENTRY.MOD
    1 Parameters    2 Sets          6 Tables    2 Activities
    0 Binaries      2 Equations     0 Logical constraints
MODLER ...D *
 PARAMETER HORIZON number of months in planning horizon
 SET t time period = 1,...,HORIZON
 SET T time period = 0,...,HORIZON
 TABLE DEMAND(t)  demand during t
 TABLE HOLDCOST(T)  holding cost during T
 TABLE PRODCOST(t)  production cost during t
 TABLE LAG1(T,t)
 TABLE EQUAL(T,t)
 TABLE SPACE(T)
 ACTIVITY I(T)  inventories goods at end of (T)
 ACTIVITY P(t)  produces goods at beginning of (t)
 EQUATION B(t)  balances inventory at end of (t)
 EQUATION COST
```

Figure 5-3. Reading An Inventory Model File

Figure 5-4 then reads in an associated data file, using the default name INVENTRY.DAT, and lists the tables LAG1 and SPACE. The table LAG1 is a control table, and SPACE is used to fix initial inventory at zero.

```
MODLER ...REA D
(confirmation messages appear)
MODLER ...L T LAG1, SPACE
  Table LAG1(T,t)      Source: INVENTRY.DAT
      ...All values = 0 EXCEPT
          T   t    Value
          -------------------
          0   1    1
          1   2    1
          2   3    1
          3   4    1
  Table SPACE(T)      Source: INVENTRY.DAT
          T     Value
          ---------------
          0        0
          1        *
          2        *
          3        *
          4        *
```

Figure 5-4. Data for Inventory Model

Figure 5-5 lists the equations. The first equation, B(t), balances material, and it is equivalent to the standard inventory equation (with no sales term). That is, B(t) = 0 is equivalent to:

$$P(t) + I(t\text{-}1) - I(t) = 0.$$

```
MODLER ...L E
   B(t) =   P(t) + LAG1(T,t)*I(T) - EQUAL(T,t)*I(T) = DEMAND(t)
   COST =   HOLDCOST(T)*I(T) + PRODCOST(t)*P(t) = MIN
```

Figure 5-5. Listing The Equations of the Inventory Model

To see this, consider first the lag term:

$$\Sigma_T \text{LAG1}(T, t)I(T),$$

where the summation is implied by T not bound to the domain of the equation, B(t). Since LAG1(T, t) = 0 for all $T \neq t-1$, and since LAG1(T, t) = 1 for T = t-1, this term reduces to I(t-1). Similarly, the last term reduces to I(t) since EQUAL(T, t) = 0 for $T \neq t$ and EQUAL(t,t) = 1.

Figure 5-6 shows the schema view. Because the activity class I appears twice in the inventory balance equation, the cell entry is multiple. To see what this means, it is necessary to see the equation listing (c.f., figure 5-5).

```
MODLER   ...  SCH D
         :::::::::: BLOCK SCHEMA ::::::::::
          I(T)          P(t)
  B(t)    LAG1/-EQUAL   1            =DEMAND
  COST    HOLDCOST      PRODCOST    ...MIN
  BOUNDS   0             0
          SPACE          *
```

Figure 5-6. Schema View of Inventory Model

There is a better way to specify control tables. See the model file INVCTRL.MOD, which came with your MODLER disk. Here the IF ... ELSE format is used to define the control tables LAG1 and EQUAL directly. They become instantiated immediately when sets T and t are instantiated. They need not be treated as data, as they are part of the model's logic. The next chapter explains an alternative to the use of control tables, namely the sum expression.

A term of an equation can be expressed in what we call the *reference form*. This is especially well suited for recursions because it does not require control tables to express the lag. The balance equation in figure 5-5 can be expressed with the following reference form.

```
B(t) =  P(t) + I{t - 1} - I{t} = DEMAND(t)
```

The curly brackets {} denote the reference form. This allows the domain specification to differ from the original domain of the activity, binding it to the equation's domain.

The inventory activity is originally defined by I(T) – that is, its domain is T = 0,...,HORIZON. The last term, I{t}, means to use t from the equation's domain, rather than the original domain, T. The middle term, I{t – 1}, also has the offset, –1.

The reference form has no more expressive power than the sum expression, which we shall consider in Chapter 6, but it is easier to read (and write). Recursions are not the only case where this is so. Further information is given in the documentation file, TABLES.DOC, and another example is in the model file,

BLENDREF.MOD. You should compare this model file with BLENDGEN.MOD, as they are different expressions of the same blending model.

Another elementary dynamic model is given in DYNLOTS.MOD, where the reference form is used. You might print it. Then, read it and its data instance into memory and write the matrix file. Use ANALYZE to examine the generated linear program.

5.3 Approximation of nonlinear functions.

Suppose the mathematical program includes a supply portion where the level of production is an increasing function of market price. In particular, consider a step function, such as in figure 5-7. The lowest step can correspond to operating a plant with existing resources with the least expensive cost of raw materials. Each step corresponds to making more supply available if consumers are willing to pay more, such as adding new capacity or purchasing more expensive raw materials. Increased costs of raw materials can happen, for example, if one must drill deeper for oil, open a new plant, or pay a tariff for an import.

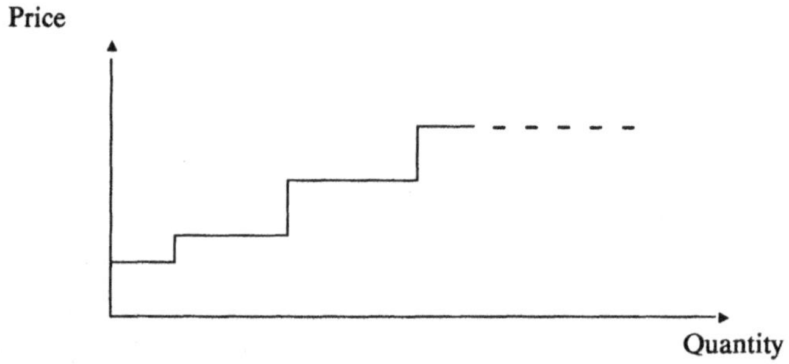

Figure 5-7. A Non-decreasing Step Function

We can now represent this in a linear program as follows. Define an activity for each step whose cost is the height of the step and whose bound is the width of the step. Due to the monotonicity, no step is active in an optimal solution unless all lower steps are at their upper bounds. One obtains, therefore, a correct approximation of the supply quantity. The supply price is precise if the solution is not degenerate – that is, if the level of the last step used is in the interior of its step. Otherwise, the supply price can be either the cost of the last step at upper bound or of the first step at zero (its lower bound).

Suppose we allow at most 10 steps, and we scale the function so that each step width must be at least 1. Then, in MODLER, this is represented as follows.

```
PARAMETERS
  NSTEPS number of steps : 0/9
SETS
  s step = 0,...,NSTEPS
TABLES
  s_cost(s)  cost     of step s : 0/*
  s_width(s) quantity of step s : 1/*
ACTIVITIES
  P(s) produce at (s) <= s_width(s)
EQUATIONS
  COST = s_cost(s)*P(s) + ...
```

This is, of course, only part of the model file. In particular, there are activities to use the supply to fulfill demands. The COST equation also includes other costs and, perhaps, revenues from sales.

A similar approach can be taken with demand functions whose quantity decreases with increasing market prices. Now the step-function is non-increasing, but the costs are negative – that is, they are revenues from sales. An example is in SUPDEMEQ.MOD.

Every non-decreasing function can be approximated with arbitrary precision with a step function, so this includes a rather broad class of supply functions. Similarly, every non-increasing function can be approximated with a step function. Therefore, monotonic supply and demand functions can be represented in a linear program in this manner.

This also highlights a limitation of MODLER, which you might have already noticed. There is no extensive data manipulation, and each term in an equation is a simple product of a data object and a decision object. The step function heights and widths must be computed by some other host system that generates the explicit declarations for MODLER (see §7.2 for some table arithmetic capability).

Another limitation is the inability to express admissibility conditions like monotonicity. In our case, we cannot require s_cost(s) < s_cost(s+1), as we would like to do in order to ensure the validity of this linear representation in each instance. Other limitations of MODLER emerge as we begin to want nonlinear modeling.

5.4 Elastic programming.

An elastic linear program is defined algebraically as:

$$\text{minimize } cx + pu + sv\text{: } y = Ax + u - v,\ L \le x, y \le U \text{ and } u, v \ge 0,$$

where u and v are called *elastic variables*. The objective coefficients, s and p, are non-negative salvage and penalty costs, respectively.

An elastic program is never infeasible, which is one reason it is a desirable model. Consider a particular level of x, and let L_y, U_y be the bounds for y. If $L_y \le Ax \le U_y$, $u=v=0$ in an optimal solution. Suppose, however, $L_y > Ax$. In this case, $u = L_y - Ax$ is a deficit to yield $y = Ax + u = L_y$ (with $v=0$). The penalty cost, p, can be regarded as a spot-market price to provide the resources necessary to make up the deficit. Making p very large corresponds to a goal of $u=0$. Now suppose $U_y < Ax$. In this case, $v = Ax - U_y$ is an excess to yield $y = Ax - v = U_y$. The salvage cost can be regarded as a loss of value. Making the salvage cost s very large corresponds to a goal of $v=0$.

Suppose a model file exists, and elastic variables are to be added to make the model an elastic program. Let E(s) be an equation with domain s, and suppose the variable names u and v have not been used. Then, if $E(s) \ge b(s)$ is the only bound (limit) specification for E, add activity v(s), and modify the equation as:

$$E(s) = \ldots - v(s),$$

where ... means the rest of the equation remains.

Similarly, if $E(s) \le b(s)$ is the only bound specification for E, add activity u(s) and modify the equation as:

$$E(s) = \ldots + u(s).$$

If the bound specification is an equality, $E(s) = b(s)$, or a range, $L(s) \le E(s) \le U(s)$, modify the equation by adding both elastic variables:

$$E(s) = \ldots + u(s) - v(s).$$

This results in at least one, and at most two, elastic variables per equation. The names u and v cannot be re-used, so some naming convention is appropriate. For example, instead of u and v in the above case, we could use uE and vE, respectively, giving a syntax: a variable that begins with u (v) is an elastic variable for the equation identified by the rest of its name. This will better support subsequent analysis with ANALYZE.

The variables must also be added to the objective equation. One can use parameters SALVAGE and PENALTY and modify the COST equation (or whatever the objective equation name is) by adding terms SALVAGE*uE(s) + PENALTY*vE(s). Alternatively, the salvage and penalty costs could be declared as one table whose domain is the set of equation classes:

```
SETS
  ...
  EQ equation
TABLES
  ...
  SALVAGE(EQ) salvage cost for (EQ)
  PENALTY(EQ) penalty cost for (EQ)
```

where ... means the other sets and tables remain.

Another alternative is to have a separate table of salvage and penalty costs for the domain of each equation:

```
TABLES
  ...
  SALV_eqn(s) salvage cost of eqn for (s)
  PENL_eqn(s) penalty cost of eqn for (s)
```

where *eqn* is to be replaced by an equation name, and *s* is to be replaced by the domain of the equation. Unless actual salvage and penalty costs are available, the simplest default is to use only the parameters.

Chapter 6. The SUM Expression

Arithmetic is where the answer is right and everything is nice and you can look out of the window and see the blue sky—or the answer is wrong and you have to start all over and try again and see how it comes out this time.

Carl Sandburg

There are situations where the implied sum is inadequate to express a linear programming model. For example, the domain of an activity class may be all regions, but its intersection with a particular row class may be with only a subset of those regions. There are many other examples where a more flexible summation is needed to express the model's equations. The SUM expression does this.

The general syntax has the form:

SUM[*index definitions* [*:index conditions*] | *term* [*: data conditions*]**]**

where **SUM[** is the key delimiter, without a space between **SUM** and **[**. The end bracket, **]**, is the delimiter that terminates the sum expression. The other brackets that are given above are metasymbols to indicate that the conditions are optional.

All of the ingredients of a sum expression are defined and illustrated in this chapter. Section 1 begins with simple sums, which have no conditions. Then, in §6.2, index conditions are defined and illustrated. The last section defines and illustrates data conditions.

The examples include those in model files that came with MODLER: SUMTEST*.MOD, which are variations of other model files that do not use sum expressions. On-line help is available with SUM.DOC.

6.1 Simple examples.

Suppose we have a term, like:

$$\Sigma_{i \in I} X_i,$$

where I is a set and X is an activity class. If the domain of X is I, we could use the implied sum with the expression, X(I). If, however, I is not the original domain of X, or if the equation fixes I, the implied sum does not work. In MODLER, this can be expressed with the sum expression:

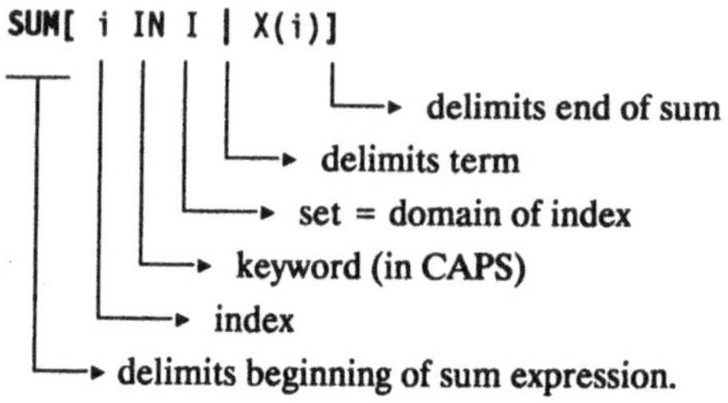

If the original domain of X is J, the sum is over $I \cap J$. If this is empty, the sum is zero.

The names of indexes can be 1 to 4 characters with no embedded delimiters. MODLER is case-sensitive, and it is good practice to define sets with capital letters and indexes with lower case letters.

The general syntax, without conditions, is given by:

SUM[*index definitions* | *term*]

Each index definition has the form:

index IN *set*

and the term has the form:

[*numeric data object* *]*decision object*

If the data object is absent, 1 is assumed, as in the previous example. Otherwise, the asterisk is mandatory to separate the data object from the decision object. The data object is a scalar (i.e., numeric constant or parameter) or a table, which could have index references in its domain. The term must contain a decision object (activity or binary), and its domain in the sum must reference each index at least once.

The decision object can also reference sets in its domain for the sum expression, and these sets need not be those in its original domain. For example, suppose the original domain of X is I×J, and we have:

```
SUM[k IN K | D(k,Z)*X(k,Z)]
```

where D is a table with domain U×V. If Z is a free set (i.e., not fixed by the equation's domain), this sum expression is equivalent to:

$$\Sigma_{k\in K\cap I\cap U,\ z\in Z\cap J\cap V} D_{kz}X_{kz}.$$

The intersections are implied by the following reasoning. An activity in class X is determined by its original domain (perhaps subject to its own data conditions). Let X_{ij} be a particular activity for $i \in I$ and $j \in J$. To see if this activity appears in the sum, we first *instantiate* k=i and z=j. If $i \notin K$, the activity does not appear in the sum. Otherwise, if $i \in K$, the term appears with coefficient D_{ij}. If either $i \notin U$ or $j \notin V$, the data object is considered to be zero, so X_{ij} is not in the sum. This reasoning yields the intersections shown.

Usually, in an expression like the above, Z=J, or $Z \subset J$ by virtue of J=Z + S. For example, J might be a set of products and Z a particular subset of the products. In addition, it is usual for the domain of D to be the same as the domain of X, or at least conformal with its domain in the sum, K×Z. These conditions are not necessary, but a well formed model will typically obey such rules of conformality – that is, set agreement.

To illustrate, figure 6-1 shows a model file, called SUMTEST1.MOD, which came with your MODLER disk (it is another expression of TRANSGEN.MOD). Let us examine the sum expressions in the equations.

```
NAME    SUMTEST1  General Transportation Model with uniform gain
PARAMETERS
 GAIN volumetric gain or loss : .9/1.1
SETS
 SR supply region
 DR demand region
 MT material
TABLES
 SUPPLY(MT,SR)         available supply of MT at SR : *, 0, 100/1000
 DEMAND(MT,DR)         required demand of MT at DR  : 0/*
 CAPACITY(MT,SR,DR) capacity for transporting MT from SR to DR :
                                                              0, 10/*
 TRANCOST(MT,SR,DR) unit cost to transport MT from SR to DR
 PRODCOST(MT,SR)     unit cost to produce MT at SR
 REVENUE(MT,DR)      unit revenue to consume MT at DR : 0, 10/100

ACTIVITIES
 P(MT,SR) produces (MT) at (SR):  PRODCOST(MT,SR) <> *
 T(MT,SR,DR) transports (MT) from (SR) to (DR) : DEMAND(MT,DR) > 0,
                    TRANCOST(MT,SR,DR) <> *, CAPACITY(MT,SR,DR) > 0
      <= CAPACITY(MT,SR,DR)
 C(MT,DR) consumes (MT) at (DR)
EQUATIONS
 S(MT,SR) limits supply of (MT) at (SR) =
          P(MT,SR) - SUM[j IN DR |T(MT,SR,j)] >= -SUPPLY(MT,SR)
 D(MT,DR) requires demand of (MT) at (DR) =
        SUM[i IN SR|GAIN*T(MT,i,DR)] - C(MT,DR) >= DEMAND(MT,DR)
 COST = SUM[m IN MT, i IN SR, j IN DR|TRANCOST(m,i,j)*T(m,i,j)] -
     REVENUE(MT,DR)*C(MT,DR) + SUM[m IN MT | PRODCOST(m,SR)*P(m,SR)]
ENDATA
```

Figure 6-1. Model File SUMTEST1.MOD

In the supply equation, `S(MT,SR)`, `MT` and `SR` are fixed sets, and the sum has one index (j), whose domain is the set `DR`. The sum expression is given by:

```
SUM[j IN DR |T(MT,SR,j)]
```

Since the original domain of T is `MT×SR×DR`, the sum expression is equivalent to the implied form, T(MT,SR,DR), as in TRANSGEN.MOD.

Suppose, however, we had defined the domain of T to be MT×SR×LO, where LO is another set of regions, perhaps a subset of DR. The appearance of T_{mij} in the sum requires: m∈MT, i∈SR and j∈LO∩DR. This is because (m, i, j) must first of all be in the original domain of T; then, the index, j, must be in its domain, which is DR.

The demand equation is similar, except the term has a parameter, GAIN. The asterisk (*), which separates the data object from the decision object, is mandatory in a sum expression; and, the sign must appear outside the sum expression.

The COST equation first has a sum over three indexes (m, i, j). The second term of COST is an implied sum, and the last sum is a mixture:

```
                          ┌bound┐
SUM[m IN MT | PRODCOST(m,SR)* P(m,SR)]
                             └─► asterisk required
```

The set SR in the domain of table PRODCOST is *bound* to that of the activity, P. Because SR is a free set in the COST equation, there is an implied sum over the supply regions. The decision object is the production activity, P. For each member of its domain, (m, i)∈MT×SR, the activity's test condition is applied. If P_{mi} is not generated, it does not appear in the sum. If it is, m and i are instantiated, and this defines the domain entry for table PRODCOST.

When the decision object has a mixture of indexes and sets in its domain for the sum expression, the sets are bound to the equation's domain, or there is an implied sum, as in a usual term. For the example in figure 6-1 we have:

```
  ┌─────────── bound ──────────┐
S(MT,SR)= ... - SUM[j IN DR |T(MT,SR,j)]
└────────── bound ─────────┘
```

In this case, the bindings mean that MT and SR are fixed sets since they comprise the domain of the equation.

Suppose the equation is instead given by:

```
  ┌─────────── bound ──────────┐
S(MT,SR)= ... - SUM[j IN DR |T(FP,SR,j)]
```

Now only the set of supply regions (SR) is bound to the equation's domain. Even if the original domain of T is MT×SR×DR, the reference to FP releases the binding to set MT in the sum expression. The result is an implied sum over the set FP.

If the data object of the term is a table, its domain is bound to that of the decision object first, then to the equation. For example, consider the following.

```
           ┌──────── bound ────────────┬──────┐
           |                           |      |
S(MT,SR)= ... - SUM[j IN DR |D(FP,j,SR)*T(FP,SR,j)]
                             └─┤bound─┘       |
                               └── bound ─────┘
```

Suppose the original domain of D is $MT \times S2 \times S3$. Then, the above sum expression means:

$$\Sigma_{k \in FP \cap MT,\, j \in S2 \cap DR}\, D_{kji} T_{kij},$$

for each $i \in SR \cap S3$.

The implied sum over $k \in FP \cap MT$ is reasoned as follows. Because the original domain of T is $MT \times SR \times DR$, the first domain reference in the sum implies $k \in MT$. Because the domain of T in the sum is $FP \times SR \times j$, $k \in FP$, so $k \in FP \cap MT$. The fact that MT is the first set in the domain of D means it has member k; it is not bound to the domain of the equation because it references FP as its first set.

The sum over $j \in S2 \cap DR$ reflects the fact that $j \notin S2$ implies $D_{kji} = 0$ since (k, j, i) is outside the domain of D, regardless of the values of k and i. That $j \in DR$ is due to the index's domain. Since $MT \times SR \times DR$ is the original domain of T, there is no further intersection to restrict j.

The restriction that $i \in S3$ is due to the domain of D. If $i \notin S3$, (k, j, i) is outside the domain of D, regardless of the values of k and j.

Here is the general rule for domain binding. First, the domain of the decision object, as referenced in the sum, instantiates all index values and the sets it references. The sets that are in the equation's domain are fixed, so the instantiation is consistent with the usual set binding. The sets that are not in the equation's domain are nevertheless instantiated, which is equivalent to an implied sum for sets that are free – that is, not fixed by the equation's domain. The table, if present as the data object of the term in the sum expression, has its own domain reference subject to the following rules. Each set reference is bound to the domain of the decision object, which has already instantiated indexes and sets. If the table's domain has additional set references, they are required to be bound to the equation's domain. Unlike a usual term (i.e., not a sum expression), there can be no free sets in the table's domain, except those referenced by the decision object. Thus, there is no implied sum for tables alone within a sum expression.

Let us consider another example, shown in figure 6-2.

```
SETS
  RM raw material
  BS blend stock
  FP final product
  MT material = BS + FP
TABLES
  SUPPLY(RM)  supply of RM
  DEMAND(FP)  demand for FP
  YIELD(RM,MT)   production yield of MT from RM
  INPUT(BS,FP)   input of BS to make FP
  OUTPUT(BS,FP)  output of FP from BS
ACTIVITIES
  P(RM,MT) produce (MT) from (RM)
  B(BS,FP) blend (BS) to make (FP)
EQUATIONS
  S(RM) = limits use of (RM) = P(RM,MT) <= SUPPLY(RM)
  B(BS) balances (BS) = SUM[r IN RM| YIELD(r,BS)*P(r,BS)] -
                        INPUT(BS,FP)*B(BS,FP)
  D(FP) demands (FP) = SUM[r IN RM| YIELD(r,FP)*P(r,FP)] +
                        OUTPUT(BS,FP)*B(BS,FP) >= DEMAND(FP)
ENDATA
```

Figure 6-2. A Blending Model

The first equation uses an implied sum to limit the use of raw material, regardless of which material is yielded. The constraint class is:

$$\Sigma_{m \in MT} P_{rm} \leq SUPPLY_r \text{ for all } r \in RM.$$

The second constraint class uses a sum expression in order to restrict the domain of the production activity and its associated yield table to the blend stocks that comprise the equation's domain. The second term is an implied sum, so the constraint class is:

$$\Sigma_{r \in RM} YIELD_{rb} P_{rb} - \Sigma_{f \in FP} INPUT_{bf} B_{bf} = 0 \text{ for all } b \in BS.$$

Finally, the demand equation uses a sum expression in order to restrict the production activity and its associated yield table to the final products that comprise the equation's domain. The second term is an implied sum, so the constraint class is:

$$\Sigma_{r \in RM} YIELD_{rf} P_{rf} - \Sigma_{b \in BS} OUTPUT_{bf} B_{bf} \geq DEMAND_f \text{ for all } f \in FP.$$

Because MT = BS + FP, the sum expression in the balance equation produces the desired constraint since BS ⊂ MT. Similarly, the sum expression in the demand equation produces the desired result since FP ⊂ MT.

6.2 Index conditions.

After the index definitions, and before the term, index conditions can be specified. The delimiter that signifies their presence is the colon (:). The general syntax is given by:

SUM[*index definitions* : *index conditions* | *term*]

Each index condition has the form:

*object*1 *relation object*2 [± *offset*]

The first object is either an index or a set map, where the index is particular to the sum. The relation is one of the six relations: <, < =, =, < >, > =, >. The second object can be literal, or it can be an index, set (possibly a map), or a numeric scalar – that is, a constant or a parameter. The comparison is performed as literals, so equals (=) and not equals (< >) are straightforward; however, the other relations could produce an anomaly if the index is not in an ordered set – that is, defined by A,...,B for scalars A and B.

There must be a blank before and after the relation (to delimit it). The offset, if present, is a scalar, and the second object must then be numeric. If the second object is an index or set, the domain for member values must be numeric. This always holds for an ordered set, but it is not necessary to have defined the set as consecutive integers. MODLER will attempt to convert the object value to numeric before adding the offset; only if this conversion fails is a fatal error issued.

The position of the set member in the comparison is irrelevant. For example, suppose i is an index with domain S1, and S2 is a set. Suppose S1 contains member 07, and S2 contains member 10. Let i be instantiated with the value 07 (where 2 characters are used for each member), and let S2 be instantiated with the value 10. The condition i < S2 is true regardless of the position of 07 in S1 and 10 in S2.

It is necessary to separate conditions with commas; and, a space is required between *index* and *relation*, before *relation*, and before and after the sign of *offset*.

Here are examples:

```
SUM[i IN SR, j IN DR: i <>  j | T(i,j)]
                      └─┴──► Space required

SUM[i IN SR, t IN T: i = DR  + 1, t > 0 | T(i,DR,t)]
                      ││  ││  │  └┴──► Space required
                      ││  ││  └──► Comma required
                      ││  └┴──► Space required
                      └┴──► Space required
```

The meaning of the first sum expression above is that the sum will be over SR×DR, except when the indexes are equal. For example, this restricts the sum of transportation activity levels to be inter-regional when SR = DR.

The second sum expression has two index conditions, and DR is a fixed set reference in the domain of the decision object, T. As we sum over i∈SR, for a particular instantiation of DR, we test if i = DR + 1. For this relation to be valid, SR and DR must have members with numeric values; otherwise, an error is issued upon instantiation of the sum expression, as when writing the matrix file. Suppose DR = {01, 02, ...} and SR = {1, 2, ...}. Let 01 be the instantiated value of DR, and let 1 be the instantiated value of i. Before the offset of 1 is added to DR, the instantiated value is converted to its numeric equivalent of 1. After adding the offset (1), the index condition becomes: 1 = 2. This is false, so the sum expression does not contain i=1 for DR=01. When i is instantiated to 2, the index condition becomes: 2 = 2. This is true, so the condition passes.

On the other hand, if the offset is not specified, there is no numeric conversion. In that case, when SR is instantiated to 1 and DR is instantiated to 01, the test 1=01 fails. Numeric conversion could be invoked by specifying an offset of 0. Thus, the condition i = DR is not equivalent to the condition i = DR + 0, because the latter forces numeric conversion of the instantiated value of DR.

The binding of DR in the first index condition is to the domain of the decision object, so DR need not be fixed by the equation's domain.

The second index condition, t > 0, requires a positive value for t whose domain is T. We could have T = 0,...,HORIZON, in which case the index condition passes for all t=1,...,HORIZON.

Figure 6-3 shows a model file, which came with your MODLER disk, that represents an inventory equation. This is equivalent to INVENTRY.MOD, which uses control tables to express the same equation.

```
NAME  SUMTEST3  Inventory problem
PARAMETERS
   HORIZON number of months in planning horizon : 1/9
SETS
   T time period = 1,...,HORIZON
TABLES
   DEMAND(T)     demand during T : 0/*
   HOLDCOST(T)   holding cost during T : 0/*
   PRODCOST(T)   production cost during T : 0/*
ACTIVITIES
   I(T) inventories goods at end of (T)
   P(T) produces goods at beginning of (T)
EQUATIONS
   COST = SUM[t IN T | HOLDCOST(t)*I(t)] + PRODCOST(T)*P(T)
   B(T) balances inventory at end of (T) = P(T) +
     SUM[t IN T: t = T - 1 | I(t)] -  I(T) = DEMAND(T)
ENDATA
```

Figure 6-3. Model File SUMTEST3.MOD

The COST equation uses a sum expression, but it could simply specify `HOLDCOST(T)*I(T)`, which is an equivalent implied sum. The balance equation, however, uses a sum expression to represent the lag term in the inventory equation.

```
                  ┌──────── bound ────────┐
B(T) = ... + SUM[t IN T: t = T - 1 | I(t)] -  I(T)
```

The binding between the set T in the equation's domain and in the index condition produces the desired result: I(t) is present in the sum for equation B(T) if, and only if, t=T–1. The resulting constraint is:

$$P_t + I_{t-1} - I_t = \text{DEMAND}_t \text{ for all } t \in T.$$

We have seen index conditions of the forms,

index relation index [± *offset*]

and

index relation set [± *offset*]

The object following the relation could also be a literal, such as the member of some set or a numeric constant. The latter was illustrated with the index condition, t > 0.

An index condition can also use set maps. Suppose we have:

set(domain) relation object [± *offset*]

where the set is a map. The domain specification in the index condition need not be the original domain. Instead, it can reference indexes or sets in the domain of the equation. In fact, it is required that every reference in the set map's domain be a set that is in the domain of the equation or a reference (set or index) in the domain of the decision object of the term.

MODLER uses the following search hierarchy for the object following the relation. First, the indexes of the sum expression are interrogated. If a match is found, the object is defined to be that index. Otherwise, the sets are interrogated next. If a match is found, the object is defined to be that set. In this case, the set must be instantiated from the domain of the decision object or of the equation, so the condition is tested with the instantiated value. Finally, if no index or set is matched, the object is presumed literal. The object value must be numeric if the offset is present; otherwise, it can be any literal up to four characters.

With these rules, an index condition is tested upon instantiation of the sum expression with the object having a specific value.

6.3 Data conditions.

After the term, data conditions can be specified. This is delimited by a colon following the term, and it has the same syntax as conditionals on activity generation:

data_object relation data_object

If a data object is a table, its domain is particular to the sum expression, so it can have a mixture of index and set references. Consider the example in SUMTEST2.MOD:

```
                                       ┌┬─ bound ─┬┐
                                       ││         ││
SUM[m IN MT | PRODCOST(m,SR)*P(m,SR):PRODCOST(m,SR) <> *]
```

The binding of the domain of the table, PRODCOST, in the data condition is to the domain of the activity class, P. Thus, the term, PRODCOST(m, SR)*P(m, SR), is in the sum only if the associated entry of PRODCOST is not infinity (indicated by the asterisk). In general, the domain specification of a table in a data condition follows the same rules as in the term. Each set reference must be bound either to the decision object or to the equation. There can be no implied summations of tables within a sum expression.

A space is required before and after each relation in the data condition. Data conditions can be compounded, as in conditionals for generation of decision objects, by separating them with commas.

Use of the sum expression is sometimes only to add readability for some modelers. Such cases have been illustrated, where the implied sum is adequate. There are other situations, however, where the expressiveness of the implied sum is limited, and a sum expression is necessary, especially to avoid complexities introduced by control tables. When viewing the model with the SCHEMA command, however, unnecessary use of the sum expression can detract from the cognition. Figure 6-4 illustrates this for two equivalent generalized transportation models, whose model files are TRANSGEN.MOD and SUMTEST1.MOD, respectively.

```
MODLER  ... READ M TRANSGEN
  (messages not shown)
MODLER  ... SCH D
              ::::::::::::::::::: BLOCK SCHEMA ::::::::::::::
              P(MT,SR)      T(MT,SR,DR)  C(MT,DR)
  S(MT,SR)    1             - 1                         >=-SUPPLY
  D(MT,DR)                  GAIN         - 1            >=DEMAND
  COST        PRODCOST      TRANCOST     -REVENUE       ...MIN
  BOUNDS      0              0           0
              *             CAPACITY     *

MODLER  ... READ M SUMTEST1
  (messages not shown)
MODLER  ... SCH D
              ::::::::::::::::::: BLOCK SCHEMA ::::::::::::::
              P(MT,SR)      T(MT,SR,DR)  C(MT,DR)
  S(MT,SR)    1             -$1                         >=-SUPPLY
  D(MT,DR)                  $GAIN        - 1            >=DEMAND
  COST        $PRODCOST     $TRANCOST    -REVENUE       ...MIN
  BOUNDS      0              0           0
              *             CAPACITY     *

  $ indicates SUM...Use LIST command to see terms of equations
```

Figure 6-4. Schema Views

The first schema view clearly shows the model's structure and data. The second schema view uses **$** to indicate sum expressions. Depending upon the individual, it may not be as immediately clear how the sum expressions limit the terms.

Chapter 7. Dependencies

The kinds of questions we ask are as many as the kinds of things which we know. They are in fact four: (1) whether the connection of an attribute with a thing is a fact, (2) what is the reason for the connection, (3) whether a thing exists, (4) what is the nature of the thing.

Aristotle

In this chapter we describe dependencies between data objects. A set can depend upon parameters or other sets; a table can depend upon parameters, sets, and other tables. Decision objects depend on data objects with domain specifications, bounds, and generation conditions. Constraints depends upon data objects, notably with domains and terms. First, the specifications of such expressions in a model file are presented in the first two sections. Then, the SHOW command is described, which shows dependency relations and illustrates how such information can be used in model management.

7.1 Set maps.

The syntax for a set map is:

set(*domain*) [*meaning*] = *range*

where the domain is a list of sets, separated by commas, which denotes the product of the sets. Following the equal sign, the range set is specified.

A set map is, therefore, dependent upon its domain sets and its range set. It is, however, an explicit set in that it requires specification in a data file (or use of the RANDOMIZ command).

On-line help for set maps is in MAPS.DOC, and an example is in the model file, JOURNEY.MOD.

7.2 Implicit sets.

The syntax for an implicit set is:

set [*meaning*] = *expression*

where the equals sign (=) delimits the presence of a dependency of this set on other objects.

The expression is one of two forms:

set union: *set* [+ ...]
numeric sequence: *scalar*,..., *scalar*

The set union expression is a sequence of sets separated by plus signs (+), which must have a blank before and after them. A single set is also possible, which simply establishes that the set being defined is equal to another set. This is useful for aliasing in equations. Here are some examples:

```
SETS
   A
   B = A
   C
   D = B + C
```

The first set, A, is primitive. Its members will be a list in a data file. Set B is defined equal to set A. Set C is primitive. Set D is the union of sets B and D. When set A is instantiated, set B will be instantiated by its dependency on A. When set C is instantiated, set D is instantiated because both B and C have their members to form their union, which is D.

The second form of expression uses two scalars to define a numeric set of consecutive integers. Each scalar is defined to be either a numeric constant or a parameter. Here are some examples:

```
PARAMETER
   HORIZON number of time periods
   UMIN least number of units
   UMAX greatest number of units
SETS
   t time period = 1,...,HORIZON
   UN unit = UMIN,...,UMAX
   Z time period = 0,...,0
   T time period = Z + t
```

The first set, t, is defined as the consecutive integers, {1, 2, ..., h}, where h is the value of the parameter, HORIZON. When HORIZON is instantiated, the set t will be instantiated with these integers. Its members will be assigned meanings:

```
1st time period
2nd time period
...
```

where "time period" is inherited from the meaning of the set.

Similarly, the set UN is instantiated when parameters UMIN and UMAX are instantiated. If the values of UMIN and UMAX are assigned a and b, respectively, the set is {a, a+1,..., b}. If it should be the case that a > b, UN is empty and MODLER issues an error message when it tries to instantiate set UN.

The set Z is instantiated immediately to be {0} since both scalars are numeric constants. The set T (distinct from t, since MODLER is case-sensitive) is the union of sets Z and t. It will be instantiated as soon as t is instantiated. It will equal t plus the member 0, which is {0, 1, ..., h}. Equivalently, you can avoid Z and define T directly, as `T = 0,...,HORIZON`.

The length of a set name determines the length of each member. In particular, the members of set t have only one character, so the parameter HORIZON is limited to the range 1/9. This should be added as a range condition to the parameter specification; otherwise, if HORIZON is instantiated with a value of 10 or more, the attempt to instantiate set t will result in an error message. If more than 9 time periods are possible in an instance, the set t should be re-named to have a second character, such as tt or tm. In that case, the first nine members are defined as 01, 02, ..., 09 (rather than 1, 2, ..., 9). The 2-character member names are expected in data files, wherever the set appears in the domain of a table. If t is re-named to have two characters, so must Z and T to be conformal. In that case, 00 is a member instead of 0.

Every implicit set is delimited by the presence of the equals sign, so the meaning of the set cannot contain that character. The two forms of implicit sets are distinguished by the ellipsis notation. The numeric set has `,...,` without embedded blanks. (The set map is distinguished by having a domain specification, indicated by the parentheses.)

7.3 Implicit tables.

The syntax for an implicit table is:

table(*domain*) [*meaning*] == *expression* [: *ranges*]

where the double equals sign (==) delimits the presence of a dependency of this table on other objects.

The expression is one of three forms:

logical:	*data_object* IF *condition* ELSE *data_object*
arithmetic:	[-] *term* [{+,-} *term* [...]]
functional:	*function*[*data_object*, *data_object*]

The logical form is distinguished by the presence of both IF and ELSE, with a blank before and after them. It is thus necessary that these words not both be used as parameter names.

In the logical form, each data object is a numeric constant, parameter or table. The table entry, associated with its domain, is assigned the first data object if the condition is true; else, it is assigned the second data object.

The condition in a logical form pertains to set members in the table's domain. Its form is:

set relation object [{+, -} *offset*] [,...]

The set is one of the sets in the table's domain, and the relation is one of the six: {<, < =, =, < >, > =, >}. The object following the relation is either a set or it is a literal, which can be a numeric constant.

The object is identified first by seeing if it is a set in the table's domain. If so, this is what is assumed. If not, the object is presumed to be a literal. Note that, as literals, 01 < 1.

The offset, if present, must be a scalar – that is, a numeric constant or a parameter. In this case, both the object and the set must be numeric. This is assured if the set is defined to be consecutive integers (with the ellipsis notation), and the object is either a numeric constant or a numeric set. It is not necessary, however, that the sets be defined this way. Upon instantiation, conversion of a set's member to a numeric value is attempted whenever the offset is present. An error message results if, and only if, this conversion attempt fails. The members can still be a list from a data file, but they must be numeric-valued.

Additional conditions can be specified, separated by commas. Here are some examples:

```
TABLES
   EQUAL(t,T) == 1 IF t = T ELSE 0
   LAG1(t,T) == 1 IF t = T - 1 ELSE 0
   FPCOST(FP)
   COST(MT) == FPCOST(FP) IF MT <> M1 ELSE 0
   A(I)
   C(I,J) == A(I) IF I = J, I <> D ELSE *
   D(T) == 1 IF T > 0 ELSE 0
```

The first table assigns the value 1 to the entries when t=T and 0 when t≠T. This appears in the model file INVCTRL.MOD to establish a control table, used in its balance equation. Table EQUAL is instantiated once sets t and T are instantiated.

Table LAG1 establishes another control table, also used in INVCTRL.MOD. Its entry is 1 if t=T-1; else, it is 0. Here the offset is -1, and both t and T must be numeric sets - that is, its members must be numeric. You might read this model file, then list its tables. You will find EQUAL and LAG1 are defined to satisfy these specifications.

Table COST is equal to the value of another table, FPCOST, when the member of its domain set, MT, is not equal to M1, which is a literal. Because the domain of FPCOST is another set, FP, not bound to the domain of COST, the sum of the entries of FPCOST becomes the table entry whenever MT <> M1. Otherwise, the table entry is 0.

Table C has two conditions for its (I, J) entry to equal A(I). First, it must be that the member of I equals the member of J, expressed as I = J. Second, it must be that the member of J is not D (literally). If these conditions are both true, the value of C(I, J) is A(I); else, its value is infinity (indicated by *).

Table D has 1 in each entry where the member of T is positive, expressed by T > 0; else, the entry is 0. This requires that the members of T be numeric, such as its definition in the previous section with the ellipsis notation.

The relations in a condition must have a blank before and after; and, if an offset is present, the + or - must also have a blank before it.

Each condition can use set maps wherever a set can appear. In that case, the domain and range of the set map must be bound to the domain of the table. For example, suppose J is a set of journeys, C is a set of cities, and ORIGIN(J) = C is a set map that specifies a city, called the origin, for each journey. Then, the following is a valid implicit table specification.

```
T(J, C) == A{J} IF ORIGIN(J) = C ELSE 0
```

Note that both the set map's domain (J) and range (C) are bound to the table's domain (note too that the reference form is used to bind the domain of A to the domain of T).

The arithmetic form of the expression is a sum, or differences, of terms. Each term has the form:

[*data_object* *]*data_object*

where the first data object is presumed 1, if absent. Otherwise, the asterisk (*) is required to separate the two data objects. If the first data object is a numeric constant or parameter, a space is required before the asterisk.

The rule of implicit sums apply to the terms: any set not bound to the domain of the table being defined implies a summation over its domain. Here are some examples:

```
TABLES
   A(I,J)
   B(I) == A(I,J) : .9/1.1
   C(J) == P
   D(I) == C(J)*A(I,J) : .9/1.1
   E(J,K) == A(I,J)*P1 - P2
   F(I,J) == B(I) + 10 *E(J,K) - A(I,J)
```

Table A is primitive (not implicit). The i-th entry of table B ($i \in I$) is the sum of A(i,J) over J. This sum is implicit because set J in the domain of A is not in the domain of B. Table B is instantiated once A is instantiated (which also implies set I is instantiated). The range condition on B is tested upon its instantiation. In this case, table B requires that each row sum of A is in the interval, [.9, 1.1].

Table C is the value of parameter P for all entries, and this is instantiated immediately upon the instantiation of P and the set J, which is its domain. This is used as a summing vector in the definition of table D. The implicit sum over J is applied to the product of C and A, and the range condition ensures that the weighted row sums of A (weighted by P) is in the interval, [.9, 1.1]. Table D could have been declared equivalently as:

```
D(I) == P *A(I,J)
```

and not use table C.

Table E has two terms. The first term is an implicit sum over set I of the product of table A and parameter P1. Then, the second term subtracts parameter P2. Since K is not referenced in the expression, each column of E will be identical upon instantiation.

The arithmetic expression for table F has three terms, which yields:

$$F(i, j) = B(i) + 10 \Sigma_{k \in K} E(j, k) - A(i, j) \;\; \forall (i, j) \in I \times J.$$

The functional form uses the square brackets, [], literally. Recognized functions are: EXPONENT, LOGARTHM, MINIMUM, MAXIMUM, BOOLE, NEGBOOLE, and SIGN (these can be abbreviated in the usual way).

EXP[D, d]	= D raised to power d (undefined for D < 0 and |d| < 1)
LOG[D, d]	= log(D: base = d) (undefined for D ≤ 0 or d ≤ 0)
MIN[D, d]	= D if D < d; else, d
MAX[D, d]	= D if D > d; else, d
BOOL[D, d]	= 1 if D = d; else, 0
NEGB[D, d]	= 0 if D = d; else, 1
SIGN[D, d]	= 1 if D > d; –1 if D < d; 0 if D = d
VALUE[S,d]	= value of member of numeric set S + d (d = offset).

In all but the last function, both data objects are numeric – that is, constants, parameters or tables. The VALUE function, however, has a numeric set as its first argument. In this case, S must be in the domain of the table.

For example, suppose R is a parameter and T(I) is a table (with domain I). Let R = 2, I = {1, 2, 3}, and T = {0, 1, 2}. Then, if F(I) = *f* [T(I), R], we have:

		F	
f	I=1	I=2	I=3
EXPONENT	0	1	4
LOGARTHM	Undefined	0	1
MINIMUM	0	1	2
MAXIMUM	2	2	2
BOOLE	0	0	1
NEGBOOLE	1	1	0
SIGN	–1	–1	0
VALUE	3	4	5

On-line help is given in the PRINT file TABLES.DOC. See, also, model files that came with your MODLER disk, especially TABLTEST.MOD.

7.4 The SHOW command.

The previous two sections have shown how sets can depend upon parameters and other sets, and how tables can depend upon parameters, sets and other tables. In addition, any object with a domain – that is, a table, decision object, or constraint – has a dependency relation on each set in its domain. Activities, binaries and logical constraints with generation conditions depend upon the parameters and tables that comprise those conditionals. The same applies to data conditions in sum expressions, where the equation then has a dependency relation with the parameters and tables that are referenced.

Dependency relations can be viewed with the SHOW command. Its syntax has the form:

```
SHOW {PARAM SET TABLE} [name]
```

where * can be used to show dependency relations for all three data objects. The name, if present, is a mask to identify the parameters, sets and tables to be shown. If no name is specified, all objects in the class specified are shown (this is the same as specifying * for the name).

Figure 1 illustrates the SHOW command with one of the model files that came with your MODLER disk, namely TRANSLGL.MOD. Table SUPPLY has one dependency: it is the upper bound for activity P. Table DEMAND is a lower bound for equation D, and so on.

Skipping to table PRODCOST, there are two dependencies. First, it appears in a conditional for the generation of activity P; second, it appears in a term of equation COST. Notice that table REVENUE has no dependencies in the model. This is identified by ** NONE, and it could signify an error, such as inadvertent omission of a term in an equation.

```
MODLER  ...SHOW TABLE
  Dependencies on Table SUPPLY(MT,SR)
    Activity P(MT,SR) <= SUPPLY(MT,SR)
  Dependencies on Table DEMAND(MT,DR)
    Equation D(MT,DR)  >= DEMAND(MT,DR)
  Dependencies on Table CAPACITY(MT,SR,DR)
    Equation T(SR,DR)  <= CAPACITY(MT,SR,DR)
  Dependencies on Table TRANCOST(MT,SR,DR)
    Equation COST : + TRANCOST(MT,SR,DR)*T(MT,SR,DR)
  Dependencies on Table PRODCOST(MT,SR)
    Activity P(MT,SR) :PRODCOST(MT,SR) <>  infinity
    Equation COST : + PRODCOST(MT,SR)*P(MT,SR)
  Dependencies on Table REVENUE(MT,DR)
     ** NONE
  Dependencies on Table CONTRACT(MT,SR,DR)
    Equation T(SR,DR)  >= CONTRACT(MT,SR,DR)
  Dependencies on Table TARGET(MT,SR,DR)
    Activity T(MT,SR,DR) >= TARGET(MT,SR,DR)
  Dependencies on Table RANGE(MT,SR,DR)
    Activity T(MT,SR,DR) <= -RANGE(MT,SR,DR)
  Dependencies on Table LINKCOST(SR,DR)
    Equation COST : + LINKCOST(SR,DR)*O(SR,DR)
```

Figure 1. Showing Dependencies on Tables in the Model File TRANSLGL.MOD

Figure 2 shows dependencies on parameters and sets in an inventory model, given in model file INVCTRL.MOD. Implied sets t and T depend on the parameter HORIZON, which defines the number of time periods in each set. Any object with t in its domain is dependent upon t, namely the tables, activity and equation shown.

```
MODLER  ...SHO P,S
  Dependencies on Parameter HORIZON
    Set t = 1,...,HORIZON
    Set T = 0,...,HORIZON
  Dependencies on Set t
    Table DEMAND(t)
    Table PRODCOST(t)
    Table EQUAL(T,t)
    Table LAG1(T,t)
    Activity P(t)
    Equation B(t)
  Dependencies on Set T
    Table HOLDCOST(T)
    Table EQUAL(T,t)
    Table LAG1(T,t)
    Table SPACE(T)
    Activity I(T)
```

Figure 2. Showing Dependencies on Parameters and Sets in the Model File INVCTRL.MOD

On-line help is available by entering HELP SHOW.

Chapter 8. File Interfaces

The fact that the great computers of today are still high-speed morons, capable of doing nothing beyond the scope of the instructions carefully programmed into them, has given many people a spurious sense of security.

Arthur C. Clarke

In this chapter we describe the types of files, which may be input or output, for MODLER. Some of these are for environment control, so we start with an explanation of special files that affect the start-up process. Then, we describe how you may enhance the on-line help, especially for others to use, with dictionaries and documents. In addition, you may prepare a sequence of MODLER commands that comprise an *execution file*. One example, which you might examine, is ATEST.EXC, which you executed just after your installation.

The only required input file to MODLER is a *model file*, which is read into memory with the READIN command. In our exercises of the previous chapters, the model files were read in, followed by a *data file*.

Most files described in this chapter are ASCII, and some are limited to record lengths of 80 bytes. On-line help is available with FILES.DOC (use the PRINT command), which is part of the reference manual generated by EXEC DOCMODL.

All of the ASCII input files recognize an INCLUDE statement. In an execution file, an INCLUDE statement allows execution of another file of commands and continuation with the main one (but the RETURN statement still returns to the terminal). In model and data files this allows some decomposition of their contents. The only restrictions are:

1. The keyword INCLUDE must begin in column 1, followed by the INCLUDE file specification.
2. Only one INCLUDE file can be active - that is, you cannot have an INCLUDE specification within another one.

The second restriction necessitates caution. For example, if you use an INCLUDE statement in an execution file that reads a model file, the model file cannot contain another INCLUDE statement.

It is recommended you setup your files to suit your environment. At the very least, examine carefully the use of execution files not only to setup the environment, but also to read model and data files into memory with an introductory message to your constituent.

8.1 Environment controls.

There is a variety of *session control commands* that help you control the environment, and you may obtain on-line help with each of these. Here we shall describe some that are relevant to files and their use.

Each file specification consists of a prefix, a filename, and a suffix. The _SETUP command allows you to change the prefix and suffix of each type. You may enter _SETUP to see the current defaults within MODLER. After installation, all prefixes are null, which means your default path is used. Each suffix (in the DOS version) corresponds to the type of file, and it too may be displayed or changed.

Before the session begins, there are special files sought by the user interface module of MODLER (called FLIP). Figure 8-1 lists these and the functions they perform. If the special file is absent, no action is taken and no message is printed.

Name	Type	Action
SETUP	EXECUTE	Change prefixes and suffixes
MODLER	DICTIONRY	Load dictionary
CREDITS	PRINT	Print
n/a	KEYWORD	Load keywords
PROFILE	EXECUTE	Execute commands

Figure 8-1. Special Files Sought Before the Interactive Session Begins

The special files are listed in order of interrogation. For example, the SETUP.EXC file may be used to set the prefix or suffix of execution files, in which case the PROFILE sought is with the new prefix and suffix. Upon installation, the PRINT suffix is .DOC, so the file CREDITS.DOC is printed. There are no KEYWORD files for MODLER, but you may add this. See the on-line help for how it can be used.

Use the PROFILE to set screen parameters (with the SCREEN command), to set strings that may be useful for reducing the number of keystrokes (with the STRING command), to set file defaults (with the _SETUP command), to load or merge dictionaries (with the _DICTNRY command), and to print an introductory message about models.

After the start-up process, the _SETUP command allows you to change prefixes and suffixes of recognized file types during the session. Figure 8-2 shows the result of entering _SETUP to obtain the current defaults.

```
MODLER ... _SETUP
                                   _SETUP Entered
      Keyword        Prefix                                Suffix
      --------------------------------------------------------
      DICTNARY                                              .DIC
      EXECUTE                                               .EXC
      HELP                                                  .HLP
      KEYWORD                                               .KEY
      PRINT                                                 .DOC
      DATA                                                  .DAT
      EXPLAIN                                               .EXP
      FORMAT                                                .FPK
      MATRIX                                                .MAT
      MODEL                                                 .MOD
      PACKED                                                .PCK
      SCHEMA                                                .SCH
      SYNTAX                                                .SYN
```

Figure 8-2. Seeing the Default File Prefixes and Suffixes with the _SETUP Command

Suppose, for example, you specify READIN MODEL TRANSGEN. The full filespec is TRANSGEN.MOD with the above defaults. But suppose you want to read in a model file whose full filespec is A:\MYPROB\TRANSGEN.MOD. Then, you may either fully qualify it as:

READIN MODEL A:\MYPROB\TRANSGEN.MOD

or you may enter _SETUP MODEL A:\MYPROB\ to change the default prefix and specify:

READIN MODEL TRANSGEN

In the following sections, the various file types are explained.

All ASCII file inputs to MODLER may have an asterisk (*) in column 1 for a comment, which is ignored.

8.2 Dictionaries, documents, help, and keyword files.

A dictionary is a file with entries of the form:

key := *text*

where *key* is any 1 to 16 character expression, and *text* is what is printed when the user asks ? *key*. The delimiter := separates the *key* from the *text*. The _SETUP file type is DICTNARY.

Examine the file MODLER.DIC, which is the dictionary that came with your MODLER system. The *text* may continue to the next line; it ends with another *key* definition on a new line or when the end of the file is reached.

To create your own dictionary you simply create an ASCII file with 80-character record lengths and enter the key and text, separated by := as shown above. MODLER does not sort the keys, so you must do that with your editor. MODLER simply treats this as a list and searches linearly for key matches. This suggests dictionaries should not be too big; they are intended to provide quick reference to a small number of terms.

A document, or PRINT file, is also created with your own editor and is presumed to be 80-character record lengths. MODLER truncates each line to fit in the screen width set by the user, so it is advisable to have left and right margins. The file is printed sequentially, so it should be short, such as those that came with MODLER (with suffix .DOC).

The HELP files are those that came with MODLER to give on-line help for each command. Generally, you will not want to change this with _SETUP during a session. If the help files are not located in the default path, you should specify this change in either SETUP.EXC or PROFILE.EXC.

The KEYWORD file type is used to support the _KEYWORD command. There are no keyword files that came with MODLER, so you can ignore this (it is used in special environments, such as to set spec files for MINOS or OB1).

8.3 Execution files.

It is sometimes desirable to prepare a sequence of MODLER commands and simply enter EXECUTE *filespec* at the MODLER prompt. The *filespec* is the EXECUTE file type recognized by the _SETUP command. The default upon installation under DOS is no prefix (i.e., your default path on your hard disk), and the suffix is .EXC (the dot is needed).

You may examine the EXECUTE files that came with MODLER by entering DIR *.EXC at the DOS prompt.

To control message spooling, there are several ways to print a comment to the screen and/or output file. In general, any ASCII input file (including the terminal) can have an asterisk in column 1 to indicate a comment. This is completely ignored; however, a dollar sign ($) in column 1 of an execution file echoes the comment to the screen. The comment can also be directed to the output file, if not the screen, by using two dollar signs ($$) in columns 1 and 2. The screen output is clipped at the screen width, but the filed output is treated like text in that it is fully printed even if more than one line is required.

See, also, the _PAUSE and _TYPE session control commands. The first issues a pause, much like the DOS PAUSE command, allowing the user to abort the execution if something is not set right. The latter types the line to the output file. Unlike the double dollar signs, the _TYPE command does not type to the screen unless that is the where output goes (controlled by the OUTPUT command) or the echo switch is on. Further, the _TYPE command has justify options (left, right and center), whereas the double dollar notation does no text formatting. Also look at the switches, such as SWITCH ECHO_ON (or simply SW ECHO), which toggles the echo. If this is on, the commands will be echoed to the screen and (if different) to the output file. This is useful for letting the user know the commands that are executed.

Sometimes it is desirable to suppress messages appearing on the screen, such as when reading a matrix file into memory. To do so, set the message switch off by specifying SWITCH MSG_ON (or simply SW M). All switches toggle, so remember to specify this again before leaving the execution file.

You might examine ATEST.EXC, which you executed just after installing MODLER. It illustrates most of the session controls you might want to have in your own execution file. One exception is the error switch, ERR_STOP. If you turn this off, the execution continues even if an error is encountered; otherwise, the execution file is aborted, returning to the screen for command input.

8.4 Model files.

The model file specifies the following objects and relations.

PARAMETERS.......to assume numeric values in specified ranges.

SETS............................basic entities, like regions, materials and time periods.

TABLES ....dimensioned by 1 or more sets, the entries will assume numeric values in specified ranges.

ACTIVITIES.............decision objects, possibly dimensioned by sets, whose values, called "levels," can be constrained by bounds.

BINARIES.................decision objects, like activities, except their values must be either 0 or 1.

EQUATIONS...........constraining or goal-directed relations, possibly dimensioned by sets, of the form, $y = Ax + Bz$, where x's are activity levels and z's are binary values, whose level (y) may be limited by bounds, $L \le y \le U$, or which may have a sense of optimization: MINIMIZE or MAXIMIZE.

LOGICALS...............constraining relations, possibly dimensioned by sets, of the form: IF antecedent THEN consequent. The antecedent and consequent are conditions, usually the level of a binary decision object.

These comprise *segments* in the model file, which must appear in the above order. In addition, there are two more (optional) segments:

BOUNDS...................defines bounds on activities (could also be defined with the activity).

LIMITS......................defines limits or criteria on equations (could also be defined with the equation).

Figure 8-3 gives complete syntax for the model file. Each *name* is 1 to 8 characters, except the set names, which we shall discuss below. The *comment* for each segment is not used, but the *meaning* given to an object or relation is part of its syntax. It is used during the session and passed on to ANALYZE via a *syntax file* (see §8.7).

```
NAME    name    [comment]
PARAMETERS [comment]
     name [meaning] [: admissible ranges]
SETS [comment]
     name [meaning] [= set [+ set [...]] | pv,...,pv]
   | name(domain) [meaning] = range set
TABLES [comment]
     name(s) [meaning] [== expression] [: admissible ranges]
ACTIVITIES [comment]
     name[(s)] [meaning] [: generation conditions]
            [bound specification]  <---- note separate line
BINARIES [comment]
     name[(s)] [meaning] [: generation conditions]
EQUATIONS [comment]
     name[(s)] [meaning] = [sign] term [sign term [...]]
     [bound specification | optspec]
LOGICALS [comment]
     name[(s)] [meaning] [: generation conditions]
            IF antecedent THEN consequent
BOUNDS [comment]
     activity boundspec
LIMITS [comment]
     equation {boundspec | optspec}
ENDATA
```

Figure 8-3. Model File Structure

Parameters and tables are data objects that can have *admissible ranges*. This is delineated by the colon (:), followed by a list of ranges. Each range is of the form: *value*[/*value*]. The first *value* is the lower value of the range. If the second *value* is specified, it is the upper value; if not, the first (and only) value is also the upper value – that is, the range degenerates to a single value.

Set names are typically 1 or 2 characters, but they can be up to 4 characters. They are used in set references – that is, domains, which we illustrate below. If a set name is k characters, all members of the set (read from a data file) must have k characters.

The equal sign (=) delineates an *implicit set*. This can be a union of sets previously defined, or it can be a set of integers denoted by the ellipsis surrounded by commas. In the latter case *pv* denotes a parameter or constant value.

The second form of a set definition is for a *set map*. Its domain is enclosed in parentheses, which is what distinguishes it syntactically from form 1. The range set follows the equal sign, which is mandatory.

The two decision objects, activities (with continuous levels) and binaries (with levels 0 or 1), have the same syntax. The conditions for generation is a list of relations of the form:

data_object relation data_object

where the *relation* is one of:

< ....less than
< =.....less than or equal to
= ....equal to
> ....greater than
> =.....greater than or equal to
< >.....not equal to.

Each data object can be a parameter, table, or constant value. Multiple conditions must be separated by commas, which can continue on the next line.

Each activity is defined simply to have non-negative levels. This can be modified by a *bounds specification*. There are three forms:

activity < data_object [AND > data_object]
activity > data_object [AND < data_object]
activity = data_object [+- data_object]

The first two forms differ only in whether an upper bound or lower bound is defined, respectively. In case both bounds are defined, the two forms differ only in which is specified first. For example,

```
P < 100 AND > 10
```

and

```
P > 10  AND < 100
```

both specify that the level of activity P must be between 10 and 100. This is inclusive, even though strict inequalities are used - that is, P < 100 is treated the same as P < = 100.

Form 3 is used to fix an activity's level, such as P = 10. It could also be used as an alternative to the first 2 forms to specify a range. For example,

P = 55 +- 45

means $10 \le P \le 100$.

This same bound specification syntax applies to the EQUATIONS section, and sets can be referenced. For example,

S(SR,MT) < SUPPLY(SR,MT)

makes equation (or activity) S(SR,MT) have an upper bound from the table SUPPLY for each member of the sets SR (say supply region) and MT (say material).

An equation can also be defined to be a criterion, or *optspec*:

```
= MINIMIZE | MAXIMIZE
```

If an equation has no bound specification and no optspec, it is presumed = 0.

The LIMITS section is also used to specify optimality criteria, which are equations whose defining value is to be minimized or maximized. Any dimensionless equation named COST is automatically assigned MINIMIZE as its optspec, and a dimensionless equation named either REVENUE or PROFIT is automatically assigned MAXIMIZE as its optspec. Other objectives must be declared, such as:

OBJ_ROW = MIN

Defaults can be removed, such as:

COST < BUDGET

where BUDGET is a parameter.

A limit can be defined immediately after an equation (next line), as in defining a bound immediately following an activity. Similarly, activity bounds can be put into the BOUNDS segment, where the syntax for bound specification is the same.

For further information, PRINT or READIN the model files that came with MODLER (suffix .MOD). PRINT DIRFILES to see a directory of the model files; and, PRINT DIRDOCS to see a directory of PRINT files, which gives additional on-line help.

8.5 Data files.

A data file specifies (part of) an instance of a model. More than one data file can be read, each with the READIN DATA specification. Objects can be read in any order, such as reading parameters after tables. Care must be taken, however, that sets are defined before tables that reference them.

The admissible ranges that are specified in the model file are checked, and a fatal error is declared if the data element violates all ranges.

The syntax for a data file is as follows. When an instance of a particular data object is defined, the first line has the form:

object_type object_name value_specification

This must begin in column 1, where *object_type* is PARAMETER, SET, or TABLE, and *object_name* is its name in the model file. The *value_specification* differs by type, which we now describe.

```
PARAMETER name = value [...]
```

assigns the numeric value to the parameter named. The [...] indicates that more than one parameter can be specified on the same line.

```
SET name [= set [+ ...]]
```

defines members for an explicit set. If the equal sign is absent, a list of members follows with a blank in column 1 of each member. The list ends when another object type is specified (in column 1) or when an end of file is reached.

Each member has a name with same number of characters as the set name. After the member name, its meaning can be given. This is not a mere comment, as it enters the syntax for subsequent analysis support with ANALYZE.

If the equal sign is present, the set is defined as a simple union of the named sets that follow, in which case the specification ends. (Continuation of sets on the next line is indicated by ending the first line with a + sign.) The equality of two sets, for example, is given by:

$$B = A$$

where A already has members. In this instance, the members of set B are those of A.

A set map differs in that it uses a relational form to list members. Each line is of the form,

domain member   range member [comment]

For example, suppose $J = \{1, 2, 3\}$ is a set of journeys, $C = \{a, b\}$ is a set of cities, and $ORIGIN(J) = C$ is a set map that assigns a city to each journey, called its origin. Then, a valid data specification is:

```
1 a
2 a
3 b
```

(column 1 is blank for each entry).

There can be missing entries. For example, it may be that only journeys 1 and 2 have an origin city. In that case, the value of ORIGIN(3) is 'undefined'. Further, unlike the simple explicit set, no meaning is given with each entry. This is because the meaning of a set map member is taken from the meaning of the range member.

If the domain contains more than one set, the domain members must appear in the same order. For example, suppose MAP(I,J) = K is a map from I×J→K. Then, an entry would appear as:

```
i j k
```

where i∈I, j∈J and k∈K. The order of the members of MAP is not relevant; in particular, there is no significance to how the domain members are listed.

To define entries for a table, specify:

```
TABLE name [= value [EXCEPT]]
```

If the equal sign is absent, the entries follow this line. There are format options, which we describe below.

If the equal sign is present, each table entry is assigned the value specified. If the EXCEPT keyword is present, exceptions to this value are listed in the *relational format*, which is described below.

When table entries follow on subsequent lines, whether it be an exhaustive list or the exceptions, the most common form is a *relational form*, defined as follows. Suppose the table's domain is the cross product of m sets, say $s_1, \ldots, s_m$. Then, each line of a relational format has the form:

```
member of s1   member of s2   ...   member of sm      value
```

The order of the sets is the same as in the table's definition in the model file. Each line contains a member of each set, then the value in the table. The domain members can be listed in any order.

If the EXCEPT form is used, only the values that differ from the value specified in the first line needs to be specified. Otherwise, the number of lines must equal the total number of table entries, which is the product of the set dimensions.

When a table has exactly 2 set references, and the size of each set is small enough, a *rectangular format* is permitted. Suppose *name*(s_1, s_2) is the table syntax, and the dimension is m×n. Then, the rectangular format is as follows.

```
                     member 1 of s2 ...          member n of s2
member 1 of s1         value11     ...              value1n
      .
      .
      .
member m of s1         valueml     ...              valuemn
```

The first line of the rectangular form lists the n members of set s_2 (in any order). Each of the next m lines begins with a member of set s_1 (in any order), followed by the n values for that row in the table. Every value must be entered in rectangular format, even zero entries.

When several tables have the same domain, it is possible to read all tables at once, using relational format. For example, suppose tables A, B and C have been defined over the domain, I. Then, the data file can specify:

```
TABLE A, B, C [= v [EXCEPT]]
* I
  1   v1 v2 v3
 etc.
```

The table names must be separated by commas. If no default value is specified, it is presumed 0, as in the case of a single table. The number of tables that can be defined at once this way is limited by the width of a line. Each entry must fit on one line. Alignments are not necessary, but it is more readable to do so.

On-line help is in FILEDAT.DOC and in RANDOMIZ.DOC (see Chapter 4 for use of the RANDOMIZ option in a data file). For further information, PRINT or READIN the data files that came with your MODLER disk (suffix .DAT), associated with respective model files.

8.6 Matrix and packed files.

A matrix file is the (de facto) standard MPS file, generally recognized by optimizers. MODLER generates this with the WRITEOUT MATRIX specification. It is also input to ANALYZE.

The packed file is an unformatted file that can be read back into MODLER with the READIN command. The reason for doing so is that it is faster than reading the model file, especially if data files have also been read.

8.7 Schema and syntax files.

The schema can be saved with SCHEMA SAVE *filespec*. This is usually not necessary because the construction for the schema display is fast, but it may be desirable to do so for exportation to a slow computer. The SCHEMA LOAD specification loads the saved schema file into memory, even if it is not for the resident model.

A syntax file defines the model syntax, which supports the English translations in ANALYZE. MODLER generates a syntax file with the WRITEOUT SYNTAX specification. The primary purpose in doing so is for the ANALYZE capabilities, but it also checks the syntax to see that the sets referenced in the syntax of a row (equation or logical) and column (activity or binary) agree with the sets that define its domain.

Upon executing the WRITEOUT SYNTAX command, MODLER checks the sets referenced in the syntax that appears in the model file with the domain of the constraint or decision object. Each set reference appears in parentheses. For example, the model file TRANSGEN.MOD contains the following syntax specification for a production activity.

```
P(MT,SR)    produces (MT) at (SR)
```

This says that the domain of activity class P is MT×SR (which were defined as sets). The syntax references these sets with the parentheses, as indicated.

Suppose, by mistake, the syntax was given as:

```
P(MT,SR)    produces (MT) at (S)
```

Then, an error is issued that set S was not found. If the model does have a set S, the error message is that S is not in the domain of P. Every set referenced in the syntax must appear in the domain of the activity (the same applies to binaries, equations and logicals).

Suppose the syntax did not reference a set in the domain. For example, suppose the syntax was given as:

```
P(MT,SR)    produces (MT)
```

In this case, a warning message is given to indicate that set SR was not referenced, but it is in the domain of P.

One can embellish the syntax specifications with the ATTRIBUT command. This assigns some attribute, like form, place or time, to a set or parameter. The purpose in doing so is to give ANALYZE more information about the syntax of the model (see chapter 9 for an example and the ANALYZE User's Guide for more examples). In addition, this enforces a conformality of types, which we now explain.

An attribute is a property that carries some meaning. For example, to attribute a set as one of form means that its members have properties of form, or mass, such as requiring units of measurement. A set of materials, for example, is a set of form and the attribution of form is a tag to inform users (human or computer) that tables having a set of form in its domain need to have units associated with them. To attribute a set as one of place means its members are locations, which do not have units of measurement, but they do have distances between them. Suppose we have a table T(MT, LO), where MT is a set of materials, attributed with form, and LO is a set of locations, attributed with place. The entries in the table are per unit of material (like tons) at a particular location. Such information can be used for model management, as in checking that units are in agreement with what they should be.

Now let us consider operationally how to attribute objects. The most primitive object in MODLER is the parameter because it cannot depend on any other object. If a parameter is attributed, all sets that depend on that parameter are attributed with the same attribute. For example, suppose HORIZON is a parameter, and we specify:

```
ATTRIBUT PARAM HORIZON = TIME
```

Then, if set T = 1,...,HORIZON, set T is attributed with the TIME attribute as well (use the SHOW command to see dependencies).

If a set is attributed, either directly by the ATTRIBUT command, or by inheritance, all sets bound to it are attributed with the same attribute. For example, suppose we specify:

```
ATTRIBUT SET LO = PLACE
```

Then, if set LO = SR, or if SR = LO, set SR is also attributed with the PLACE attribute. More generally, if LO is bound to sets by LO = S1 + S2 + ... + Sn, or by S1 = LO + S2 + ... + Sn, then sets S1, ..., Sn are all attributed with the PLACE attribute.

Multiple inheritance is not allowed. Each parameter and set can have at most one attribute; and, all bindings enforce a conformality of attributes. Once a parameter or set is attributed, all of its bound parameters and sets are attributed with the same attribute instantly. An attempt to attribute a set that is already attributed, such as S1 above, results in an error message.

MODLER defines the attributes of FORM, PLACE and TIME, but one can specify any additional attribute. Each new attribute is added to the list if its name does not conflict with those already defined. For example, an attribute of F is rejected because its name conflicts with FORM as a telescopic abbreviation. The attribute name must not be a valid (i.e., telescopic) abbreviation of any attribute already defined. Specifying ATTRIBUT with no further specification results in listing the attributes.

8.8 Explain files.

When preparing a model for someone else to use, it is useful to have a tailored explanation of the model. When using MODLER, the EXPLAIN command prints the explain file. Figure 8-4 illustrates this with a blending model (which is discussed in §9.3).

MODLER ... READ M BLEND
MODLER ... EXPLAIN

This model represents a single plant that contains primary and blend units. The following diagram shows that each primary unit uses raw feed stocks as input and produces blend stocks and final products as output. Each blend unit uses blend stocks as input and produces final products.

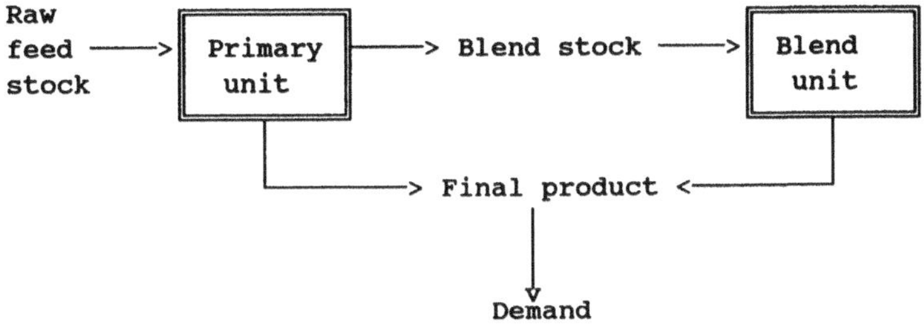

The BLEND model defines sets PU, BU, RF, BS and FP to represent primary units, blend units, raw feed stock, blend stock and final products, respectively.

There are two classes of activities, OP and OB, which operate primary and blend units, respectively_ The objective is to minimize the total cost of operating units, subject to demand requirements (for final products)_ Here are the equations_

Total cost (objective):

```
COST = COST_PU(PU,RF)*OP(PU,RF)     +      COST_BU(BU)*OB(BU)
       ------------------------            ------------------
       summed over primary units           summed over blend units,
       (and the raw feed stocks),          this is total cost to
       this is total cost to operate       operate blend units.
       primary units.
```

Figure 8-4. Using the EXPLAIN Command (continued on next page)

```
Supply limit of raw feed stocks:
                                                              same
                                                              raw feed
                                                              stock
SUP(RF) = INPUT-RF(PU,RF)*OP(PU,RF) <= SUPPLY(RF)

          summed over primary units

Total capacity limit of primary units:
                                                    same
                                                    primary unit
CAP(PU) = OP(PU,RF) <= CAP-PU(PU)

          summed over
          raw feed stock

Balance blend stocks:                                              same
                                                                   blend stock
BAL(BS) = OUTPT-BS(PU,BS)*OP(PU,RF)  -  INPUT-BS(BU,BS)*OB(BU) = 0

          summed over primary units,    summed over blend units,
          this is total blend stock     this is total blend stock
          produced                      consumed

Demand requirements:                                                 same
                                                                     final product
DEM(FP) = OUTPT-PU(PU,FP,RF)*OP(PU,RF) + OUTPT-BU(BU,FP)*OB(BU)

          summed over primary units,     summed over blend units,
          this is total products         this is total products
          they produce                   they produce

 >= DEMAND(FP)

    demand requirement
    for same final product
```

Figure 8-4. Using the EXPLAIN Command (continued from previous page)

If an explain file is not prepared (with the model's name), MODLER gives a default explanation, shown in figure 8-5.

```
To understand the model, use LIST MODEL and SCHEMA DISPLAY commands.
Then, use LIST and DISPLAY commands to see more about the objects.
To get a sense of the model's behavior, read data or use the
RANDOMIZ command to generate instances.  Then, use ANALYZE to see
the details of the instance.
```

Figure 8-5. Default Explain File

Chapter 9. More Model Description

How can it be that mathematics, being after all a product of human thought independent of experience, is so admirably adapted to the objects of reality?
Albert Einstein

This chapter goes further into illustrating the use of MODLER for model description and instance generation. First, we review the three types of term expressions in equations. Then, we look at different expressions of a blending model. The expressions we consider are equivalent, and our purpose is to illustrate advantages of each form of expression. In §9.3, we describe the REPORT execution file, which contains MODLER commands to report on a model description. It can be changed to suit your environment, and it can be used as a batch file by an INCLUDE in your PROFILE, ending with a QUIT command. In §9.4, the ATTRIBUT command is used to assign attributes to parameters and sets, with inheritance by implied sets and tables. Then, generation conditions for equations are illustrated with a critical path scheduling model. Finally, set maps are illustrated.

9.1 Term expressions.

Let us first review three forms of expressions of the terms that comprise an equation. An example of the simple form is:

```
S(MT,SR) = P(MT,SR) - T(MT,SR,DR)
D(MT,DR) =            T(MT,SR,DR)
```

The same equations can be given with sum expressions:

```
S(MT,SR) = P(MT,SR) - SUM[d IN DR | T(MT,SR,d)]
D(MT,DR) =            SUM[s IN SR | T(MT,s,DR)]
```

The only advantage of the sum expression in this case is to remind ourselves that we are summing over the free sets (DR in the first equation, and SR in the second). More generally, the sum expression gives greater expressive power, namely when there are conditions in the sum (i.e., index relations with the sets in the equation's domain or data conditions to delineate terms for an instance). On the other hand, when the added expressive power is not needed, the simple form is more readable once we are used to the notion of an implied summation over free sets – that is, sets not fixed by the domain of the equation.

The third form, called the reference form, is designed for readability when the full expressive power of the sum is not needed but when the simple form is inadequate. In our example, suppose the domains of the equations use a common set of locations, say LO, and the transportation activity is defined over aliases: T(MT, SR, DR), where SR = LO and DR = LO in the model description. In this case, the simple form cannot be used because it requires the original domain. Suppose we attempt to write:

```
S(MT,LO) = P(MT,LO) - T(MT,SR,DR)
D(MT,LO) =            T(MT,SR,DR)
```

Then, there is no binding between SR and LO in the first equation, and there is no binding between DR and LO in the second equation. In this case, we obtain an incorrect formulation because the transportation activities will be summed over its supply and demand regions. The algebraic equivalent of the above is:

$$S(m, s) = P(m, s) - \Sigma_{i,j}\, T(m, i, j) \text{ and } D(m, d) = \Sigma_{i,j}\, T(m, i, j)$$

for $m \in$ MT, $s \in$ LO, $d \in$ LO, where the summations are over all $(i, j) \in$ SR×DR.

The sum expression overcomes this by binding SR and DR in the transportation activity to the domain of the equation. It is only the binding that is needed, and the reference form allows this without a sum expression. In this example, we can write:

```
S(MT,LO) = P(MT,LO) - T{MT,LO,DR}
D(MT,LO) =            T{MT,SR,LO}
```

The use of the brackets, T{...}, instead of the parentheses, T(...), delimits the reference form. It says that there is a domain deviation: each set is either the original set in the domain of the activity, or it is a set in the domain of the equation. In the first equation, T{MT,LO,DR} references the domain MT×LO×DR, where LO is a deviation from its original set, SR. Instead, the reference is bound to the set LO in the equation's domain.

```
                 Bound
   ┌───────────────┬───────────────┐
S(MT,LO)  =  P(MT,LO)  -  T{MT,LO,DR}
      └───────────────┴───────────────┘
                 Bound
```

Besides the simple domain deviation of using a set from the equation, the reference form allows offsets. For example, a standard inventory recursion could be written as:

```
B(MT,T) = I{MT,T-1} + P(MT,T) - S(MT,T) - I{MT,T},
```

where the original domain of I is MT×Z for Z = 0,...,HORIZON, while T = 1,...,HORIZON. The first inventory activity has a lag of one period (T-1), which is in its domain for all T. The last inventory activity specifies T, rather than its original domain, Z, which binds it to the time period in the balance equation's domain.

In the inventory equation, the simple form cannot be used, except with control tables (c.f., §5.2). The sum expression can be used, but it obscures the meaning of the equation (c.f., chapter 6). The reference form has the readability of the simple form with the expressive power of the sum expression (in this case).

In the next section we compare three expressions of the same blending model. This will help to clarify the differences among the three forms.

9.2 A blending model.

In this section we consider three model files that express the same blending model differently. Figure 9-1 shows the equations section of a blending model file (sans the comments), called BLENDGEN.MOD, which came with your MODLER system. Let us discuss each equation.

The COST equation uses only the simple form, where each term is an implied summation over the domains of the activities and data tables.

The supply equation, S(RF, R, T), fixes sets RF, R and T, so the first term is simply the associated production activity, P(RF, R, T). The second term is a sum expression, where the materials in the domain of the inventory activity (I) is restricted to raw feed stock (RF). This restriction is declared by the index condition, m = RF, binding it to the equation's domain, which includes the set RF. Further, the time period is lagged, by the index condition, t = T − 1. To see precisely how this works, consider an activity, I(m, r, t), where (m, r, t) ∈ MT×R×Z ≡ domain of I.

To see if I(m, r, z) is in equation S(i, j, k) for (i, j, k) ∈ RF×R×T ≡ domain of equation S, this term is instantiated with m, r and t in sets MT, R and Z, respectively. The index condition m = RF, defines i = m, and the index condition t = T − 1 defines k = z + 1. The relation, r = j, is from the binding of set R in both the domain of I and of S. Thus, I(m, r, z) appears with a +1 coefficient in equation S(m, r, z+1) for m ∈ RF, r ∈ R and z ∈ {0,...,HORIZON−1}.

The second sum expression behaves similarly, where indexes m and t are bound to sets RF and T, respectively, in the equation's domain. It is necessary to specify t as an index in both sum expressions in order to bind the time set of the inventory activity to the equation's domain, with the proper meaning: entering inventory minus ending inventory.

```
COST = COST_BUY(MT,R)*P(MT,R,T)      + COST_O&M(UN,R)*O(UN,R,T) +
       COST_TRN(MT,S,D)*T(MT,S,D,T) + COST_INV(MT,R)*I(MT,R,Z)
S(RF,R,T) limits use of (RF) in (R) at (T) =
       P(RF,R,T) +
       SUM[m IN RF, t IN Z : m = RF, t = T - 1 | I(m,R,t)] -
       SUM[m IN RF, t IN Z : m = RF, t = T | I(m,R,t)] -
       SUM[u IN PU | INPUT_RF(u,RF)*O(u,R,T)] +
       SUM[m IN RF, s IN S, d IN D: m = RF, d = R | T(m,s,d,T)] -
       SUM[m IN RF, s IN S, d IN D: m = RF, s = R  | T(m,s,d,T)]
    >= 0
B(BS,R,T) balances (BS) in (R) at (T) =
       SUM[m IN BS, t IN Z: m = BS, t = T - 1 | I(m,R,t)] -
       SUM[m IN BS, t IN Z: m = BS, t = T  | I(m,R,t)] +
       SUM[u IN PU | OUTPT_BS(u,BS)*O(u,R,T)] -
       SUM[u IN BU | INPUT_BS(u,BS)*O(u,R,T)] -
       SUM[m IN BS, s IN S, d IN D: m = BS, s = R | T(m,s,d,T)] +
       SUM[m IN BS, s IN S, d IN D: m = BS, d = R | T(m,s,d,T)]
D(FP,R,T) demands (FP) at (R) at (T) =
       SUM[m IN FP, t IN Z : m = FP, t = T - 1 | I(m,R,t)] -
       SUM[m IN FP, t IN Z : m = FP, t = T | I(m,R,t)] +
       SUM[u IN PU | OUTPT_FP(u,FP)*O(u,R,T)] +
       SUM[u IN BU | OUTPT_FB(u,FP)*O(u,R,T)] -
       SUM[m IN FP, s IN S, d IN D: m = FP, s = R | T(m,s,d,T)] +
       SUM[m IN FP, s IN S, d IN D: m = FP, d = R | T(m,s,d,T)]
    >= DEMAND(FP,R,T)
```

Figure 9-1. The Equations Section of the Model File, BLENDGEN.MOD

The next sum expression is to restrict summing over primary units, since the operation activity, O(UN,R,T), has all units in its domain. There are no index conditions because there is no special binding to the equation's domain (other than the region, R, and the time period, T, which is bound simply). If we had defined the domain of the operation activity to be only for primary units, we could use a simple form, like INPUT_RF(PU,RF)*O(PU,R,T). But there is only one class of process operation activity in this description, so the domain of O is UN×R×T, not PU×R×T. We could also re-define table INPUT_RF to have domain UN×RF and let the data entry specify zeroes for blending units. This model description, however, specifies the domain of table INPUT_RF as PU×RF.

Skipping to the transportation activities, the terms for in-bound transportation minus out-bound transportation are expressed by the sum expressions:

```
- SUM[m IN BS, s IN S, d IN D: m = BS, s = R | T(m,s,d,T)] +
  SUM[m IN BS, s IN S, d IN D: m = BS, d = R | T(m,s,d,T)]
```

bound to equation's domain

Figure 9-2 shows the portion of the model file, BLENDREF.MOD (which also came with MODLER), that differs from the BLENDGEN model file, namely the equations section.

```
COST = COST_BUY(MT,R)*P(MT,R,T)      + COST_O&M(UN,R)*O(UN,R,T) +
       COST_TRN(MT,S,D)*T(MT,S,D,T) + COST_INV(MT,R)*I(MT,R,Z)
S(RF,R,T) limits use of (RF) in (R) at (T) = P(RF,R,T) +
       I{RF,R,T - 1} - I{RF,R,T} + T{RF,S,R,T} - T{RF,R,D,T} -
       SUM[u IN PU | INPUT_RF(u,RF)*O(u,R,T)]  >= 0
B(BS,R,T) balances (BS) in (R) at (T) =
       I{BS,R,T - 1} - I{BS,R,T} + T{BS,S,R,T} - T{BS,R,D,T} +
       SUM[u IN PU | OUTPT_BS(u,BS)*O(u,R,T)] -
       SUM[u IN BU | INPUT_BS(u,BS)*O(u,R,T)]
D(FP,R,T) demands (FP) at (R) at (T) =
       I{FP,R,T - 1} - I{FP,R,T} + T{FP,S,R,T} - T{FP,R,D,T} +
       SUM[u IN PU | OUTPT_FP(u,FP)*O(u,R,T)] +
       SUM[u IN BU | OUTPT_FB(u,FP)*O(u,R,T)]
    >= DEMAND(FP,R,T)
```

Figure 9-2. The Equations Section of the Model File, BLENDREF.MOD

The first difference occurs with the inventory terms. The reference form, used in figure 9-2, expresses the entering inventory minus the ending inventory simply, as:

```
I{RF,R,T - 1} - I{RF,R,T},
```

where the domains referenced, RF×R×(T–1) and RF×R×T, respectively, are bound to the equation's domain. The first term has an offset of –1, which is the lag time for carrying inventory in this model. These two terms express the same difference as the sum expression, but with greater clarity.

The difference between in-bound and out-bound transportation is also expressed with a reference form:

This is equivalent to the sum expressions in 9-1. The form is

```
T{RF,S,R,T} - T{RF,R,D,T},
```

where the referenced domain projections, RF×R×T, are bound the equation's domain. The free sets, S in the first term and D in the second, are interpreted as implied summations. The algebraic equivalent, for (f, r, t) ∈ RF×R×T (≡ equation's domain), is:

$$\Sigma_{k\in S}\, T(f, k, r, t) - \Sigma_{k\in D}\, T(f, r, k, t).$$

The reference form is needed (rather than simple form) to deviate from the original domain, MT×S×R×T, in order to reference the proper material (RF) and region (R), which are bound to the equation's domain. The same simplifications in readability apply to the balance equation for blend stocks and to the demand equation for final products.

Figure 9-3 shows an alternative, where we can eliminate sum expressions altogether by re-defining the domains of the input and output tables.

```
COST = COST_BUY(MT,R)*P(MT,R,T)       + COST_O&M(UN,R)*O(UN,R,T) +
       COST_TRN(MT,S,D)*T(MT,S,D,T) + COST_INV(MT,R)*I(MT,R,Z)
S(RF,R,T) limits use of (RF) in (R) at (T) = P(RF,R,T) +
       I{RF,R,T - 1} - I{RF,R,T} + T{RF,S,R,T} - T{RF,R,D,T} -
       INPUT_RF(UN,RF)*O(UN,R,T) >= 0
B(BS,R,T) balances (BS) in (R) at (T) =
       I{BS,R,T - 1} - I{BS,R,T} +
       OUTPT_BS(UN,BS)*O(UN,R,T) - INPUT_BS(UN,BS)*O(UN,R,T) +
       T{BS,S,R,T} - T{BS,R,D,T}
D(FP,R,T) demands (FP) at (R) at (T) =
       I{FP,R,T - 1} - I{FP,R,T} + T{FP,S,R,T} - T{FP,R,D,T} +
       OUTPT_FP(UN,FP)*O(UN,R,T)
    >= DEMAND(FP,R,T)
```

Figure 9-3. The Equations Section of the Model File, BLENDALT.MOD

Consider the simple form in the supply equation,

```
INPUT_RF(UN,RF)*O(UN,R,T)
```

The implied summation over process units (UN) applies because table INPUT_RF now has domain UN×RF, so UN is bound to the activity's domain and RF is bound to the equation's domain.

For model descriptions in BLENDGEN.MOD and BLENDREF.MOD, the domain of this table is PU×RF, so PU was not bound to UN. In this case, however, each instance must specify that the entries in table INPUT_RF are zero, except for the appropriate primary process units. This is easy to do since MODLER allows a supersparse data specification (see §8.5), but it might be more desirable to specify the correct domain of INPUT_RF in the model description.

The three model descriptions differ in their readability not only in the listing of their equations, but also in other views of the model. In particular, consider the way schema are displayed for the three formulations.

Sum expressions are indicated by the dollar sign ($), and domain deviations (i.e., reference form) are indicated by the colon (:). These are shown for the three forms in the following three figures.

```
                  :::::::::::::::::::::: BLOCK SCHEMA ::::::::::::::::::::::
                  P(RF,R,T)    O(UN,R,T)    T(MT,S,D,T)  I(MT,R,Z)
COST              COST_BUY     COST_O&M     COST_TRN     COST_INV      ...MIN
S(RF,R,T)          1           -$INPUT_RF   $1/-$1       $1/-$1        >= 0
B(BS,R,T)                      Multiple     -$1/$1       $1/-$1         = 0
D(FP,R,T)                      Multiple     -$1/$1       $1/-$1        >=DEMAND
BOUNDS             0            0            0           INV_LO
                  MAX_BUY      CAPACITY      *           INV_UP

$ indicates SUM...Multiple indicates term objects cannot fit in cell
...Use LIST command to see terms of equations
```

Figure 9-4. Schema for BLENDGEN

```
            :::::::::::::::::::::: BLOCK SCHEMA ::::::::::::::::::::::::::
            P(RF,R,T)    O(UN,R,T)    T(MT,S,D,T)  I(MT,R,Z)
COST        COST_BUY     COST_O&M     COST_TRN     COST_INV     ...MIN
S(RF,R,T)    1           -$INPUT_RF   :1/-:1       :1/-:1       >= 0
B(BS,R,T)                Multiple     :1/-:1       :1/-:1        = 0
D(FP,R,T)                Multiple     :1/-:1       :1/-:1       >=DEMAND
BOUNDS      0             0           0            INV_LO
            MAX_BUY      CAPACITY     *            INV_UP

$ indicates SUM and : indicates domain deviation...Multiple indicates term
objects cannot fit in cell...Use LIST command to see terms of equations
```

Figure 9-5. Schema for BLENDREF

```
            :::::::::::::::::::::: BLOCK SCHEMA ::::::::::::::::::::::::::
            P(RF,R,T)    O(UN,R,T)    T(MT,S,D,T)  I(MT,R,Z)
COST        COST_BUY     COST_O&M     COST_TRN     COST_INV     ...MIN
S(RF,R,T)    1           -INPUT_RF    :1/-:1       :1/-:1       >= 0
B(BS,R,T)                Multiple     :1/-:1       :1/-:1        = 0
D(FP,R,T)                OUTPT_FP     :1/-:1       :1/-:1       >=DEMAND
BOUNDS      0             0           0            INV_LO
            MAX_BUY      CAPACITY     *            INV_UP

: indicates domain deviation...Multiple indicates term objects cannot fit
in cell...Use LIST command to see terms of equations
```

Figure 9-6. Schema for BLENDALT

Now consider the relative merits of the three formulations. The simple form cannot express such things as deviations from the original domain of the decision object, including lag terms on time sets. The sum expression offers the most expressive power and can accommodate bindings to the equation's domain. For some, the sum expression is also very readable, so nothing else is needed. For others, the reference form offers a more readable expression than the sum, and it can accommodate certain deviations from the simple form. In this case, the trade-off is cognitive understanding, which we call readability, with expressive power. All three forms have their place, and the example illustrates their differences.

9.3 The REPORT execution file.

Figure 9-7 shows an execution file, which came with MODLER, called REPORT.EXC. It is a generic report writer that gives the model description from a variety of views. The first four lines are simply comments (indicated by the asterisk in column 1). Let us apply this to the blending model, BLENDALT.MOD (c.f., figure 9-3).

```
* File:     REPORT.EXC
* Date:     8-24-91
* Author:   Harvey J. Greenberg
* Function: Execute file for MODLER to give report to OUTPUT file

 SUMMARY
$$   :::::::::::::::::::::::::::::::::::::::::::::::::::::::::::::::::::
$$
 LIST MODEL
$$
 SCHEMA DISP
$$   :::::::::::::::::::: OBJECT DESCRIPTIONS :::::::::::::::::::::::::
 DISP * //E
$$   :::::::::::::::: DEPENDENCY INFORMATION :::::::::::::::::::::::::
$$                          PARAMETERS
$$                          ==========
 SHO PARAM
$$
$$                            SETS
$$                            ====
 SHO SET
$$
$$                           TABLES
$$                           ======
 SHO TABLE
$$  ::::::::::::::::::::::::: END REPORT :::::::::::::::::::::::::::::
* RETURN EXCLUDED, SO THIS CAN BE AN INCLUDE FILE
```

Figure 9-7. The REPORT.EXC File

Figure 9-8 illustrates a sequence that begins with reading the BLENDALT model, with the base case data that was read when ATEST was executed (after installation). Before executing the report generator, we set the screen length to virtual infinity to suppress page ejection, and we reduce its width to 75 columns.

Then, we spool the output to BLENDALT.DOC. Figure 9-8 shows what appears on your screen. Those lines with the $$ in columns 1 and 2 are comments that are written to the screen (with $) and to the output file (without $). The rest of this section describes the output in detail. The output file, BLENDALT.DOC, is too large to put into one figure, so we present it in several figures. You can try this yourself and look at BLENDALT.DOC as we go through its contents.

```
MODLER  ...REA P BLENDALT
 READIN complete
MODLER  ...SCR W 75 L 9999999
 Screen WIDTH=        75 LENGTH=      32000
MODLER  ...O BLENDALT.DOC
MODLER  ...EXEC REPORT
 $   :::::::::::::::::::::::::::::::::::::::::::::::::::::::::::::::::
 $
 $
 $   ::::::::::::::::::  OBJECT DESCRIPTIONS  ::::::::::::::::::::::::
 $   ::::::::::::::::  DEPENDENCY INFORMATION  :::::::::::::::::::::::
 $                          PARAMETERS
 $                          ==========
 $
 $                            SETS
 $                            ====
 $
 $                           TABLES
 $                           ======
 $  ::::::::::::::::::::::  END REPORT  ::::::::::::::::::::::::::::
 RETURN...back to terminal
MODLER  ...O *
MODLER  ...SCR W=79 L=23
```

Figure 9-8. Executing REPORT to Document a Blending Model

Referring to figure 9-7, the first command gives a summary of the model. Then, a blank line, followed by a string of colons, is written to both the screen and to the output file (indicated by $$ in columns 1 and 2). This separates the SUMMARY output from the model listing, as shown in figure 9-9. Then, we see a blank line, followed by the string of colons, where were specified by the $$ in the report execution file. The model is then listed.

```
        Summary of model BLENDALT ...General Blending Model
Source file:  BLENDALT.MOD
  3 Parameters   12 Sets        19 Tables     4 Activities
  0 Binaries      4 Equations    0 Logical constraints

:::::::::::::::::::::::::::::::::::::::::::::::::::::::::::::::::::

                           Model BLENDALT
                       General Blending Model
Minimize COST
  Subject to:

 COST =  COST_BUY(RF,R)*P(RF,R,T) + COST_O&M(UN,R)*O(UN,R,T)
       + COST_TRN(MT,S,D)*T(MT,S,D,T) + COST_INV(MT,R)*I(MT,R,Z)
 S(RF,R,T) =  P(RF,R,T) + I(RF,R,T-1) - I(RF,R,T) + T(RF,S,R,T)
            - T(RF,R,D,T) - INPUT_RF(UN,RF)*O(UN,R,T) >= 0
 B(BS,R,T) =  I(BS,R,T-1) - I(BS,R,T) + OUTPT_BS(UN,BS)*O(UN,R,T)
            - INPUT_BS(UN,BS)*O(UN,R,T) + T(BS,S,R,T) - T(BS,R,D,T) = 0
 D(FP,R,T) =  I(FP,R,T-1) - I(FP,R,T) + T(FP,S,R,T) - T(FP,R,D,T)
            + OUTPT_FP(UN,FP)*O(UN,R,T) >= DEMAND(FP,R,T)
  Decision variables:
                       0 <= P <= MAX_BUY
                       0 <= O <= CAPACITY
                       0 <= T
                  INV_LO <= I <= INV_UP
```

Figure 9-9. The SUMMARY and LIST MODEL Output of the Report

In figure 9-10, the output of the report continues with the block schema. The first four column strips are the activity classes, shown with their original domains (see BLENDALT.MOD for the original model description). The first row strip is the COST equation, which is the minimand. Compare this view with the equation listed in figure 9-9. Note that the simple form is used for each term, so the schema simply shows the data object that is the coefficient. For example, the first term in the equation is COST_BUY(RF,R)*P(RF,R,T), so the table COST_BUY appears in the schema cell defined by the COST row strip and the P(RF,R,T) column strip.

```
                ::::::::::::::::::::::: BLOCK SCHEMA :::::::::::::::::::::::
                P(RF,R,T)    O(UN,R,T)    T(MT,S,D,T) I(MT,R,Z)
COST            COST_BUY     COST_O&M     COST_TRN     COST_INV     ...MIN
S(RF,R,T)       1            -INPUT_RF    :1/-:1       :1/-:1       >= 0
B(BS,R,T)                    Multiple     :1/-:1       :1/-:1        = 0
D(FP,R,T)                    OUTPT_FP     :1/-:1       :1/-:1       >=DEMAND
BOUNDS          0             0           0            INV_LO
                MAX_BUY      CAPACITY     *            INV_UP

: indicates domain deviation...Multiple indicates term objects cannot fit
in cell...Use LIST command to see terms of equations
```

Figure 9-10. Block Schema From Report Execution

The second row strip is the supply equation, S(RF,R,T). Each production activity, P(RF,R,T), has a unity coefficient with the same domain member, so a 1 appears in the schema cell. Compare this with the equation listing in figure 9-9. The last two cells before the right-hand side are :1/–:1. The slash is used to separate multiple entries of the same activity class in an equation. From figure 9-9, note that the transportation activity class, whose name is T with domain MT×S×D×T, appears in one term for transportation into region R of the material (MT) and in another term for transportation out of region R. As stated in the schema's footnote, the colon in each term representation indicates a domain deviation. This represents the reference form: T{RF,S,R,T} – T{RF,R,D,T}. In each term there are two deviations from the original domain of T. First, the material is raw feed (RF). Second, the first term, which is in-bound transportation of raw feed into region R, references region R for the destination region, and the second term, which is out-bound transportation, references region R for the source region.

A similar deviation applies to the inventory terms, I{RF,R,T – 1} – I({RF,R,T}. In both cases, the material is restricted to RF. The first term has an offset (–1) to the equation's time period (T), which is a lag term that describes the inventory at the end of the previous period.

In the balance equation for blend stocks, the first cell is blank because there the production activity does not appear in the equation. The next cell entry is Multiple, which means the class of operation activities appears in more than one term, and the string length of the icon OUTPT_BS/–INPUT_BS is greater than what could fit in a cell. MODLER sets cell widths based on the string lengths of the row and column strips and on the number of column strips. It also uses a maximum cell length of 16 characters, which this entry exceeds. The intention is to try to fit the schema display in rectangular form for a screen, whose width is generally 80 columns. If the display does not fit, the output is given in relational format, which is not as visual as the rectangular form (try the SCHEMA DISPLAY with a screen width of only 65 to see the relational output).

In figure 9-11, the object descriptions are given from the DISPLAY command, which specified extended format (by //E in figure 9-7). The first line, which is a header, is the **$$** comment line in REPORT.EXC (c.f., figure 9-7). The display begins with the number of parameters stated (3 in this model), followed by the (extended) display of each parameter.

```
 :::::::::::::::::: OBJECT DESCRIPTIONS ::::::::::::::::::::::::
Parameter NUM_PU...number of primary units
         Admissible values:   1/10
         Current value:         5.000   from BLENDALT.DAT
Parameter NUM_BU...number of units
         Admissible values:   11/20
         Current value:        15.000   from BLENDALT.DAT
Parameter HORIZON...planning horizon
         Admissible values:   1/9
         Current value:         3.000   from BLENDALT.DAT
```

Figure 9-11. Parameters Displayed in Extended Format

The first line of each parameter displayed is its name and meaning. The next line gives its admissible values. For parameter NUM_PU, the first one displayed, its value is restricted to the interval, [1, 10]. Line 3 of each parameter display is its current value and source. Parameter NUM_PU currently has the value 5, which was specified in the data file, BLENDALT.DAT.

After the three parameters are displayed, the sets follow, shown in figure 9-12. Each set's display gives its number of members, its name and its meaning on one line, and its source on a second line.

```
Set PU has 5 members...primary unit
    = 1,...,NUM_PU
Set BU has 5 members...blend unit
    = 11,...,NUM_BU
Set UN has 10 members...process unit
    = PU + BU
Set RF has 2 members...raw feed stock
    From BLENDALT.DAT
Set BS has 4 members...blend stock
    From BLENDALT.DAT
Set FP has 4 members...final product
    From BLENDALT.DAT
Set MT has 10 members...material
    = RF + BS + FP
Set R has 3 members...process region
    From BLENDALT.DAT
Set S has 3 members...source region
    = R
Set D has 3 members...destination region
    = R
Set T has 3 members...time period
    = 1,...,HORIZON
Set Z has 4 members...time period
    = 0,...,HORIZON
```

Figure 9-12. Sets Displayed in Extended Format

In figure 9-13, the display of objects continues with tables. Those shown first are the explicit tables only because that is the order in which they are defined in the model file. For each table, the first line gives its name, domain and meaning. The second line gives its admissible values (which may continue with more lines, if necessary). These are the admissible ranges defined in the model file. Finally, the source of current entries is given. For an explicit table, the source is the name of the data file (BLENDALT.DAT in this case).

```
Table COST_O&M(UN,R) ...cost to operate UN in R
         Admissible values:      0/*
         Current values from BLENDALT.DAT
Table RATE(MT) ...cost rate ($/mile) for MT
         Admissible values:      0/*
         Current values from BLENDALT.DAT
Table DISTANCE(S,D) ...distance (miles) from S to D
         Admissible values:      0/*
         Current values from BLENDALT.DAT
Table COST_TRN(MT,S,D) ...cost to transport MT from S to D
         Admissible values:  any
         Current values from = RATE(MT)*DISTANCE(S,D)
Table COST_BUY(RF,R) ...cost to buy RF
         Admissible values:      0/*
         Current values from BLENDALT.DAT
Table COST_INV(MT,R) ...cost to hold inventory of MT in R
         Admissible values:  any
         Current values from BLENDALT.DAT
Table CAPACITY(UN,R) ...capacity of UN in R
         Admissible values:      0,  10/500
         Current values from BLENDALT.DAT
Table INPUT_RF(UN,RF) ...amount of RF used by UN
         Admissible values:      0/1
         Current values from BLENDALT.DAT
Table INPUT_BS(UN,BS) ...amount of BS used by UN
         Admissible values:      0/1
         Current values from BLENDALT.DAT
Table OUTPT_BS(UN,BS) ...amount of BS produced by UN
         Admissible values:      0/1
         Current values from BLENDALT.DAT
Table OUTPT_FP(UN,FP) ...amount of FP produced by UN
         Admissible values:      0/1
         Current values from BLENDALT.DAT
Table SUPPLY(MT,R) ...initial supply of MT in R
         Admissible values:      0/*
         Current values from BLENDALT.DAT
Table DEMAND(FP,R,T) ...demand for FP in R
         Admissible values:      0,  10/*
         Current values from BLENDALT.DAT
```

Figure 9-13. Continuation of Display with Explicit Tables

In figure 9-14, the display of objects continues with the remaining tables. The form of the display is the same, with the source of implicit tables being its description in the model file.

The first implicit table displayed is CHK_INV(MT, R) = INV_MAX(MT, R) − INV_MIN(MT, R). With the condition 0/* (i.e., entries must be in $[0, \infty)$), this checks that the upper bound is greater than the associated lower bound for the inventory activities.

The implicit table INV_LO(MT, R, Z) is the actual lower bound on the associated inventory activity, I(MT, R, Z). The "adjustment" is that it is the data table value of INV_MIN for positive time periods, but it is the initial supply, SUPPLY(MT, R), for the initial time period, indicated by the condition Z = 0. A similar situation applies to the inventory upper bounds, INV_UP(MT, R, Z).

```
Table MAX_BUY(RF,R) ...maximum amount of RF that can be purchased in R
          Admissible values:      0/*
          Current values from BLENDALT.DAT
Table INV_MIN(MT,R) ...minimum stock level
          Admissible values:      0,  10/1000
          Current values from BLENDALT.DAT
Table INV_MAX(MT,R) ...maximum stock level
          Admissible values:      0/*
          Current values from BLENDALT.DAT
Table CHK_INV(MT,R) ...checks inventory range
          Admissible values:      0/*
          Current values from =INV_MAX(MT,R) - INV_MIN(MT,R)
Table INV_LO(MT,R,Z) ...adjusted min stock level
          Admissible values:  any
          Current values from = SUPPLY(MT,R) IF Z = 0 ELSE INV_MIN(MT,R)
Table INV_UP(MT,R,Z) ...adjusted max stock level
          Admissible values:  any
          Current values from = SUPPLY(MT,R) IF Z = 0 ELSE INV_MAX(MT,R)
```

Figure 9-14. Continuation of Display with Implicit Tables

In figure 9-15, the display of objects continues with the activity classes. The first line gives the name, domain and meaning of the activity. The second line displays generation conditions, and the third line displays the bounds of the activity. Each of the activity classes uses an associated cost table to determine the generation of members of its class. For example, the purchase activity P(f, r, t), for (f, r, t) ∈ RF×R×T, is generated if, and only if, COST_BUY(f, r) is not infinity.

```
Activity P(RF,R,T) ...purchases (RF) in (R) at (T)
         When:   COST_BUY(RF,R) <  infinity,
                 MAX_BUY(RF,R) >  0
         Bounds:   >=  0  AND <=  MAX_BUY(RF,R)
Activity O(UN,R,T) ...operates (UN) in (R) at (T)
         When:   COST_O&M(UN,R) <  infinity,
                 CAPACITY(UN,R) >  0
         Bounds:   >=  0  AND <=  CAPACITY(UN,R)
Activity T(MT,S,D,T) ...transports (MT) from (S) to (D) at (T)
         When:   COST_TRN(MT,S,D) <  infinity
         Bounds:   >=  0 (no upper bound)
Activity I(MT,R,Z) ...inventories (MT) in (R) at end of (Z)
         When:   COST_INV(MT,R) <  infinity,
                 INV_UP(MT,R,Z) >  0
         Bounds:   >=  INV_LO(MT,R,Z)  AND <=  INV_UP(MT,R,Z)
```

Figure 9-15. Display of Activity Decision Objects

Figure 9-16 shows the output from the last portion of the DISPLAY command, which are the equations. (There are no binary decision objects and no logical constraints in this model.) The display of the equation classes gives their names, domain and meaning on the first line, and their limit specification on the second line. The COST equation is a minimand, so = MIN appears. These could be eliminated from the report, particularly since the LIST MODEL command already lists the equations. Thus, the REPORT.EXC execution file could qualify the display command to: DISPLAY PARAM,SET,TABLE (or simply D P,S,T). This would limit the objects of display to parameters, sets and tables.

```
Equation COST ...
         = MIN
Equation S(RF,R,T) ...limits use of (RF) in (R) at (T)
         >=  0
Equation B(BS,R,T) ...balances (BS) in (R) at (T)
         =   0
Equation D(FP,R,T) ...demands (FP) at (R) at (T)
         >=  DEMAND(FP,R,T)
```

Figure 9-16. Final Portion of Display Output

In figure 9-17, we begin the output from the SHOW command, following the headers (generated by the $$ comment in REPORT.EXC), with the dependencies on each parameter. The set PU depends upon the parameter, NUM_PU, since the set is a numeric sequence from 1 through the number of primary units. Similarly, set BU depends upon parameter NUM_BU. Both sets T and Z depend upon the HORIZON parameter. (The blank line, after showing parameter dependencies, is from the $$ comment line.)

```
    :::::::::::::::  DEPENDENCY INFORMATION  :::::::::::::::::::::::::
                           PARAMETERS
                           ==========
Dependencies on Parameter NUM_PU
  Set PU = 1,...,NUM_PU
Dependencies on Parameter NUM_BU
  Set BU = 11,...,NUM_BU
Dependencies on Parameter HORIZON
  Set T = 1,...,HORIZON
  Set Z = 0,...,HORIZON
```

Figure 9-17. Beginning of Showing Dependency Information with Sets That Depend Upon the Parameters

In figure 9-18, the first portion of the set dependencies are shown, namely for the sets of units. The set UN depends upon both PU and BU since it is defined to be their union (UN = PU + BU). The dependencies on set UN include all objects that contain set UN in its domain, namely the six tables and the one activity shown.

```
                              SETS
                              ====
Dependencies on Set PU
  Set UN contains PU
Dependencies on Set BU
  Set UN contains BU
Dependencies on Set UN
  Table COST_O&M(UN,R)
  Table CAPACITY(UN,R)
  Table INPUT_RF(UN,RF)
  Table INPUT_BS(UN,BS)
  Table OUTPT_BS(UN,BS)
  Table OUTPT_FP(UN,FP)
  Activity O(UN,R,T)
```

Figure 9-18. Showing Dependencies on Sets of Processing Units

In figure 9-19, the output of the SHOW command continues with sets of materials, RF, BS, FP and MT. The set MT depends upon sets RF, BS and FP since it is their union (MT = RF + BS + FP). The tables, activities and equations that are shown in the dependency list of a set include those for which that set is in their domain.

```
Dependencies on Set RF
  Set MT contains RF
  Table COST_BUY(RF,R)
  Table INPUT_RF(UN,RF)
  Table MAX_BUY(RF,R)
  Activity P(RF,R,T)
  Equation S(RF,R,T)
Dependencies on Set BS
  Set MT contains BS
  Table INPUT_BS(UN,BS)
  Table OUTPT_BS(UN,BS)
  Equation B(BS,R,T)
Dependencies on Set FP
  Set MT contains FP
  Table OUTPT_FP(UN,FP)
  Table DEMAND(FP,R,T)
  Equation D(FP,R,T)
Dependencies on Set MT
  Table RATE(MT)
  Table COST_TRN(MT,S,D)
  Table COST_INV(MT,R)
  Table SUPPLY(MT,R)
  Table INV_MIN(MT,R)
  Table INV_MAX(MT,R)
  Table CHK_INV(MT,R)
  Table INV_LO(MT,R,Z)
  Table INV_UP(MT,R,Z)
  Activity T(MT,S,D,T)
  Activity I(MT,R,Z)
```

Figure 9-19. Showing Dependencies on Sets of Materials

In figure 9-20, more sets are shown, namely the sets of location: R, S and D. Sets S and D depend upon R since they are defined (in BLENDALT.MOD) to be equal to R. They serve as aliases in defining some tables and in defining the domain of the transportation activity class.

```
Dependencies on Set R
  Set S = R
  Set D = R
  Table COST_O&M(UN,R)
  Table COST_BUY(RF,R)
  Table COST_INV(MT,R)
  Table CAPACITY(UN,R)
  Table SUPPLY(MT,R)
  Table DEMAND(FP,R,T)
  Table MAX_BUY(RF,R)
  Table INV_MIN(MT,R)
  Table INV_MAX(MT,R)
  Table CHK_INV(MT,R)
  Table INV_LO(MT,R,Z)
  Table INV_UP(MT,R,Z)
  Activity P(RF,R,T)
  Activity O(UN,R,T)
  Activity I(MT,R,Z)
  Equation S(RF,R,T)
  Equation B(BS,R,T)
  Equation D(FP,R,T)
Dependencies on Set S
  Table DISTANCE(S,D)
  Table COST_TRN(MT,S,D)
  Activity T(MT,S,D,T)
Dependencies on Set D
  Table DISTANCE(S,D)
  Table COST_TRN(MT,S,D)
  Activity T(MT,S,D,T)
```

Figure 9-20. Showing Dependencies on Sets of Regions

In figure 9-21, the last portion of sets are shown, which are sets of time periods: T and Z. Most temporal objects depend on T, which recall is the set of time periods, T = 1,...,HORIZON. The inventory activity class and associated bound tables, INV_LO and INV_UP, are defined for the initial time period as well, where Z = 0,...,HORIZON.

```
Dependencies on Set T
  Table DEMAND(FP,R,T)
  Activity P(RF,R,T)
  Activity O(UN,R,T)
  Activity T(MT,S,D,T)
  Equation S(RF,R,T)
  Equation B(BS,R,T)
  Equation D(FP,R,T)
Dependencies on Set Z
  Table INV_LO(MT,R,Z)
  Table INV_UP(MT,R,Z)
  Activity I(MT,R,Z)
```

Figure 9-21. Showing Dependencies on Sets of Time Periods

In general, parameters and sets observe the following dependency rules.

1. Set S depends upon parameter P if S is an implicit numeric set, and P appears in its definition.

2. Set S depends upon set A if S is an implicit set defined as the union of sets that include A.

In figure 9-22, the first portion of tables are shown, following the header that was generated by the $$ comment lines in REPORT.EXC (c.f., figure 9-7). A dependency on a table is established with an activity that references the table in its generation conditions. Thus, activity O(UN,R,T) is dependent upon table COST_O&M(UN,R) because the members of UN×R×T for which an operation activity is generated are those for which the operation and maintenance cost is not infinite (see BLENDALT.MOD). This is a way to specify which process units are available in each region, which we assume remains constant over time. The second dependency is with the COST equation because table COST_O&M appears in a term, namely the coefficient of the associated operation activity.

```
                          TABLES
                          ======
Dependencies on Table COST_O&M(UN,R)
  Activity O(UN,R,T) :COST_O&M(UN,R) <   infinity
  Equation COST : + COST_O&M(UN,R)*O(UN,R,T)
Dependencies on Table RATE(MT)
  Table COST_TRN(MT,S,D)  = RATE(MT)*DISTANCE(S,D)
Dependencies on Table DISTANCE(S,D)
  Table COST_TRN(MT,S,D)  = RATE(MT)*DISTANCE(S,D)
Dependencies on Table COST_TRN(MT,S,D)
  Activity T(MT,S,D,T) :COST_TRN(MT,S,D) <   infinity
  Equation COST : + COST_TRN(MT,S,D)*T(MT,S,D,T)
Dependencies on Table COST_BUY(RF,R)
  Activity P(RF,R,T) :COST_BUY(RF,R) <   infinity
  Equation COST : + COST_BUY(RF,R)*P(RF,R,T)
Dependencies on Table COST_INV(MT,R)
  Activity I(MT,R,Z) :COST_INV(MT,R) <   infinity
  Equation COST : + COST_INV(MT,R)*I(MT,R,Z)
Dependencies on Table CAPACITY(UN,R)
  Activity O(UN,R,T) <= CAPACITY(UN,R)
                     :CAPACITY(UN,R) >   0
```

Figure 9-22. First Portion of Showing Dependencies on Tables from SHOW Command

The second table, RATE(MT), appears in the definition of the implicit table, COST_TRN(MT,S,D), which is the transportation cost of the material (MT) from a source region (S) to a destination region (D). Thus, the entries in table COST_TRN depend upon the entries in table RATE. Similarly, COST_TRN depends upon the table, DISTANCE(S,D), which is the distance from S to D. We observe the following dependency rules.

1. A table depends upon table T if it is an implicit table and T appears in its definition.
2. An activity depends upon table T if T appears in any of its generation conditions, or if T appears in its bound specification.
3. An equation depends upon table T if T appears in a term of the equation, or if T appears in its limit specification.

These rules extend to binary decision objects and to logical constraints. The dependency relation is that of an immediate neighbor in a dependency graph. That is, an equation may depend upon table T, and T may depend upon another table, S. The equation appears in the dependency list of only table T, not in the dependency list of table S.

In figure 9-23, the dependencies for the remaining tables are shown. The last statement, which indicates the end of the report, was generated by the **$$** comment line in REPORT.EXC (c.f., figure 9-7).

```
Dependencies on Table INPUT_RF(UN,RF)
  Equation S(RF,R,T) : - INPUT_RF(UN,RF)*O(UN,R,T)
Dependencies on Table INPUT_BS(UN,BS)
  Equation B(BS,R,T) : - INPUT_BS(UN,BS)*O(UN,R,T)
Dependencies on Table OUTPT_BS(UN,BS)
  Equation B(BS,R,T) : + OUTPT_BS(UN,BS)*O(UN,R,T)
Dependencies on Table OUTPT_FP(UN,FP)
  Equation D(FP,R,T) : + OUTPT_FP(UN,FP)*O(UN,R,T)
Dependencies on Table SUPPLY(MT,R)
  Table INV_LO(MT,R,Z)  = SUPPLY(MT,R) IF Z = 0 ELSE INV_MIN(MT,R)
  Table INV_UP(MT,R,Z)  = SUPPLY(MT,R) IF Z = 0 ELSE INV_MAX(MT,R)
Dependencies on Table DEMAND(FP,R,T)
  Equation D(FP,R,T)  >= DEMAND(FP,R,T)
Dependencies on Table MAX_BUY(RF,R)
  Activity P(RF,R,T) <= MAX_BUY(RF,R)
                      :MAX_BUY(RF,R) >   0
Dependencies on Table INV_MIN(MT,R)
  Table CHK_INV(MT,R)  =INV_MAX(MT,R) - INV_MIN(MT,R)
  Table INV_LO(MT,R,Z)  = SUPPLY(MT,R) IF Z = 0 ELSE INV_MIN(MT,R)
Dependencies on Table INV_MAX(MT,R)
  Table CHK_INV(MT,R)  =INV_MAX(MT,R) - INV_MIN(MT,R)
  Table INV_UP(MT,R,Z)  = SUPPLY(MT,R) IF Z = 0 ELSE INV_MAX(MT,R)
Dependencies on Table CHK_INV(MT,R)
  ** NONE
Dependencies on Table INV_LO(MT,R,Z)
  Activity I(MT,R,Z) >= INV_LO(MT,R,Z)
Dependencies on Table INV_UP(MT,R,Z)
  Activity I(MT,R,Z) <= INV_UP(MT,R,Z)
                      :INV_UP(MT,R,Z) >   0
  ::::::::::::::::::::::  END REPORT  ::::::::::::::::::::::::::::
```

Figure 9-23. Last Portion of Dependencies on Tables from SHOW Command

Note that no other object depends upon table CHK_INV(MT,R), which is indicated in the output by ** NONE. Examination of the model file reveals this is all right, as table CHK_INV is just a check on the inventory bounds. It does not appear in the description of any constraint (equation or logical) or in any decision object (activity or binary). There are other times, however, when the appearance of no dependency is an error. It could be that the table should have appeared somewhere, or that the table is erroneously defined, perhaps carried over from an earlier model description.

9.4 Attribute inheritance.

Let us now see the effect of the ATTRIBUT command. Figure 9-24 shows the screen messages when we apply attribute the property of time to the parameter, HORIZON. The sets, T and Z, which depend upon HORIZON inherit this attribute, as indicated by the subsequent display. Note that the DISPLAY command delineated the parameters to be only those that have been attributed. This is specified in the option field, following the extended format specification (//E).

```
MODLER  ...ATTR PARAM HORIZON = TIME
 Set T    inherits attribute TIME
 Set Z    inherits attribute TIME
          3 Attributed
MODLER  ...D P //E ATTR = *
  Parameter HORIZON...planning horizon.  Attribute: TIME
           Admissible values:   1/9
           Current value:         3.000   from BLENDALT.DAT
```

Figure 9-24. Assigning the Time Attribute to Parameter HORIZON

Now ATTRIBUT PLACE to any of the sets of place, R, S or D, and note the other two inherit this attribute, as shown in figure 9-25. This is because S = R and D = R in the model description. Any attribute assigned to a set is inherited by all sets that share a dependency with it. Once S is attributed as a set of place, set R inherits this attribute. In turn, this causes D to inherit the attribute of place.

```
MODLER  ...ATTR SET S = PLACE
 Subset R    inherits attribute PLACE
 Set D    inherits attribute PLACE
         3 Attributed
MODLER  ...D S //E A = PLACE
  Set R has 3 members...process region.  Attribute: PLACE
      From BLENDALT.DAT
  Set S has 3 members...source region.  Attribute: PLACE
      = R
  Set D has 3 members...destination region.  Attribute: PLACE
      = R
```

Figure 9-25. Assigning the Place Attribute to Set S

9.5 Equation conditions.

In all of the previous models, there was no need to restrict the generation of equations. Now we illustrate such a case with a critical path scheduling model.

We are given a set of tasks. For each task, we are given the normal time it takes to complete it and the cost under normal conditions. Some tasks can be done in crash time at a higher cost. In the construction industry, examples of tasks are: site preparation, laying foundation, decking, framing, putting up exterior walls, roofing, and exterior finishing. Some tasks cannot begin until other tasks are completed. For example, the foundation cannot be laid until site preparation is completed, and framing cannot begin until the foundation is laid. These are called *precedence constraints.*

Suppose $S_i \equiv$ start time of task i, and $T_i \equiv$ time to complete task i. Then, a precedence constraint has the form:

$$S_i \geq S_p + T_p,$$

where task p is a predecessor of task i. This says that task i cannot begin before task p is completed.

The time to complete a task is its normal time unless it can be performed under crash conditions. In that case, let C_i be the decision variable that compresses completion time, where $0 \leq C_i \leq d_i \equiv$ normal time – crash time to complete task i. Then, the precedence constraint becomes:

$$S_i - S_p + C_p \geq N_p,$$

where $N_p \equiv$ normal time to complete task p.

If $C_p = 0$, the constraint requires the start time of task i to be after the start time of task p plus the normal time it takes to complete task p. The start time of task i can be decreased if C_p becomes positive. If $C_p = d_p$, its maximum amount of compression, the precedence constraint becomes:

$$S_i - S_p \geq N_p - d_p = \text{crash time to complete task p.}$$

The cost to complete all tasks is the simple sum of task costs under normal conditions, which is constant since all tasks must be completed, plus the additional cost of crashing on chosen tasks. Ignoring the constant, the variable cost, due to crashing, is therefore $\Sigma_i\ c_iC_i$, where $c_i \equiv$ crash cost – normal cost to complete task i. In addition, there can be a penalty cost for lateness.

To compute lateness, we first need the finishing time, say F. This is the time it takes to complete all tasks, so $F \geq S_i + N_i - C_i$. In words, this inequality says that the finishing time must be at least as great as the start time of task i plus its completion time, which is $N_i - C_i$. Further, this is true for all tasks. Now suppose the scheduled time to finish all tasks is T, and let L denote the lateness. If $F \leq T$, $L = 0$; and, if $F > T$, $L = F - T$. This can be represented by $L \geq F - T$ if we also require $L \geq 0$ and have a positive penalty cost, say p*L.

Putting together these model objects and relations, we arrive at the model shown in figures 9-26 and 9-27 (the first two lines are comments that give the reference for the model formulation). Most of the specifications are straightforward expressions of what we just described, but there are several things about this MODLER representation that require discussion.

There are three parameters: InfTime, OnTime and PENALTY. These control which objective to minimize, which we describe below. There is only one set of tasks (TA), but two aliases are defined: PR and ta (note lower case, which makes ta distinct from TA). In addition, there is a dummy set (d), which is implicit, with only one entry. The need for these implicit sets pertains to the way MODLER establishes bindings in equations, which we shall describe shortly.

First, notice the reference form used to define table LEAST. The original domain of table NORMAL_T is the set TA, but we want LEAST(PR) to be the normal time of completion for a predecessor task for its appearance in the precedence constraints (equation P).

```
LEAST(PR) == NORML_T{PR} IF PRED(ta,PR) >= 1 ELSE 0
       |                 |                |
       |_________________|________________|

  Set PR is bound to domain of table LEAST
```

```
* R.M. Stark and R.H. Mayer, Jr., Quantitative Construction Management,
*       Wiley, New York, NY, 1983.

NAME CRITPATH Critical Path for Scheduling
PARAMETERS
 InfTime infinite time : 99
 OnTime  time scheduled to finish all tasks : 1/9999
 PENALTY penalty cost for lateness : 0/*
SETS
 TA task
 PR predecessor task = TA
* THE FOLLOWING IS TO RELEASE TA BINDING IN EQUATION P
 ta task = TA
 d  dummy = 1,...,1
TABLES
 NORML_C(TA) cost of TA in $000 under normal conditions
 CRASH_C(TA) cost of TA in $000 under crash conditions
 DIFF_C(TA) differential cost of TA in $000 ==
       CRASH_C(TA) - NORML_C(TA) : 0/*
 NORML_T(TA) completion time of TA in weeks under normal conditions
 CRASH_T(TA) completion time of TA in weeks under crash conditions
 DIFF_T(TA) differential time of TA == NORML_T(TA) - CRASH_T(TA) : 0/*
 PRED(ta,PR) = 1 if PR is predecessor of ta : 0, 1
 LEAST(PR) normal time if PR is a predecessor of some task ==
       NORML_T(PR) IF PRED(ta,PR) >= 1 ELSE 0
 MUSTFNSH(d) == * IF PENALTY > 0 ELSE OnTime
```

Figure 9-26. First Portion of a Critical Path Scheduling Model (CRITPATH.MOD)

This says that the value of LEAST(p), for $p \in PR$, equals the normal completion time if task p is a predecessor of some task. This condition is established because of the implied sum in the condition, PRED(ta, PR) > = 1, over the set ta. That is, this condition is true if, and only if, PRED(i, p) = 1 for some $i \in ta$ for the particular task, p.

Table MUSTFNSH is the time all tasks must finish. Its dummy domain set (d) is because MODLER requires each table to have at least one set in its domain. (This is really a parameter, but MODLER does not permit implicit parameters.) The condition specifies that there is no constraint – that is, the value of MUSTFNSH is infinity (indicated by *) if PENALTY > 0. The value of PENALTY is positive only if the objective is the total cost, including a penalty for lateness. Otherwise, if PENALTY = 0, no constraint is imposed upon the finishing time. In particular, this applies if we seek to minimize finishing time using only normal times (and costs). We shall return to this point below when we say more about the multiple objectives.

```
ACTIVITIES
 S(TA) starts (TA)
 C(TA) crashes (TA) : CRASH_C(TA) <> *, OnTime < InfTime
        <= 1
 Finish finishes all tasks
        <= MUSTFNSH(d)
 Late   allows finish time to be late
EQUATIONS
 P(ta,PR) requires (ta) to end before (PR) can begin =
        S(ta) - S(PR) + DIFF_T(PR)*C(PR)
        >= LEAST(PR) : PRED(ta,PR) = 1
 F(TA) requires finish time to finish (TA) = Finish - S(TA) +
        DIFF_T(TA)*C(TA) >= NORML_T(TA)
 LATE  defines lateness of finish time = -Late + Finish
        <= OnTime : OnTime < InfTime

* Objectives : generation conditions for well posed
*            (only one objective will be written to matrix file)
 COST totals crash costs  = DIFF_C(TA)*C(TA) = MINIMIZE : OnTime < InfTime
 TIME defines finish time = Finish = MINIMIZE : OnTime >= InfTime
 TCOST totals crash and lateness costs = DIFF_C(TA)*C(TA) +
        PENALTY*Late = MINIMIZE : PENALTY > 0
ENDATA
```

Figure 9-27. Second Portion of a Critical Path Scheduling Model (CRITPATH.MOD)

Second, notice that set ta is simply an alias of TA, the set of tasks. It is needed for the domain of the precedence equation in order to release binding with TA. Let us see why it is invalid to use what appears to be the more natural expression:

```
P(TA,PR) = S(TA) - S(PR) + C(PR) >= LEAST(PR)
```

Consider the generation of coefficients for S_i for $i \in TA$. Member i of set TA is instantiated, and the reference form in the second term binds it to PR. This causes the term to appear in equation P_{ii}, which is incorrect. By having the alias, ta, appear in the equation's domain, there is no binding to TA.

Third, the precedence constraints have generation conditions: P(ta, PR) = 1. This means equation P_{ij} is generated only if task j is a predecessor of task i, which is indicated by the table entry, PRED(i, j). Similarly, equation LATE is generated only if the condition, OnTime < InfTime is true. This condition is a scenario control, where parameter OnTime is used to define lateness.

Finally, consider the three objectives. In any instance, one of these is the generated objective, depending upon the equation conditions. Figure 9-28 shows the data file for minimizing completion time under normal conditions.

```
* Data file for CRITPATH.MOD to minimize completion time
* Example from same reference.
* SOLUTION:  MIN TIME = 17
PARAMETER OnTime = 999, PENALTY = 0
ENDATA
```

Figure 9-28. Data File to Minimize Completion Time (CPMTIME.DAT)

To generate this instance, perform the following commands.

```
READ MOD CRITPATH            ←— read model file
READ DAT                     ←— read base data file (CRITPATH.DAT)
READ DAT CPMTIME             ←— read CPMTIME.DAT
WRIT MAT CPMTIME //TIME      ←— write matrix file with objective = TIME
```

If you examine the base data file (CRITPATH.DAT), you will see the set of tasks and associated data defined. The model file specified parameter InfTime = 99, so we have, for this instance, OnTime > InfTime, which means we do not want to constrain finishing time.

Now let us review what gets generated for this instance. Activity C(TA) does not get generated for any task because the generation condition, OnTime < InfTime, fails. For the same reason, equation COST is not generated. Table MUSTFNSH = * in this case, so no upper bound is generated for activity Finish; and, activity Late is not generated. Equation TIME is generated, and its value is to be minimized. Equation TCOST is not generated because PENALTY = 0.

Thus, the instance for CPMTIME is as follows.

MINIMIZE TIME subject to:

TIME = Finish

$P_{ip} = S_i - S_p \geq \text{NORMAL\_T}(p)$ for i, p : PRED(i, p) = 1

$F_i = \text{Finish} - S_i \geq \text{NORMAL\_T}(i)\ \forall i$

$S_i \geq 0\ \forall\, i,\ \text{Finish} \geq 0.$

Figure 9-29 shows the data file to generate an instance that minimizes differential crash cost. Now, since OnTime < InfTime, the crash activities (C) are generated, as are activity Late and equation COST. Equation TCOST is still not generated since PENALTY = 0.

```
* Data file for CRITPATH.MOD
* This scenario is to minimize differential costs to finish
* tasks before MIN TIME solution (17 weeks), which uses only
* normal time.  This solution must use crash time.
* SOLUTION:  MIN (DIFFERENTIAL) COST = 7 WITH CDE=CIF=1 AS CRASHES
PARAMETER OnTime = 14, PENALTY = 0
ENDATA
```

Figure 9-29. Data File to Minimize Differential Crash Cost (CPMCRASH.DAT)

The instance generated for CPMCRASH is as follows.

MINIMIZE COST subject to:

$COST = \Sigma_i \, DIFF\_C(i) C_i$

$P_{ip} = \quad S_i - S_p \geq NORMAL\_T(p)$ for i, p : PRED(i, p) = 1

$F_i = \quad Finish - S_i + DIFF\_T(i) C_i \geq NORMAL\_T(i) \; \forall i$

$Late = -Late + Finish \leq 14$

$S_i, C_i \geq 0 \; \forall \, i, \; Finish, Late \geq 0.$

The last instance is to minimize total costs, including a penalty for lateness, which is equation TCOST. Figure 9-30 gives the associated data file.

```
* Data file for CRITPATH.MOD
* Example similar to CPMCRASH.DAT, except lateness penalty added.
* SOLUTION:  MIN TOTAL COST = 6.4
*        WITH CRASH ON ONLY INTERIOR FINISHING (CIF = 1)
*        AND Late = 2 weeks.  The lateness penalty displaced crashing
*        on the foundation (this will be true for PENALTY < 2.5).
*        For PENALTY < 2, the MIN TIME solution is obtained with
*        total cost (TCOST) = PENALTY*3 (Late = 3 weeks).
PARAMETER OnTime = 14, PENALTY = 2.2
ENDATA
```

Figure 9-30. Data File to Minimize Total Crash and Lateness Costs (CPMTCOST.DAT)

You can read this model and data into memory and query it. You might also generate its matrix files (for the three cases indicated above) and examine the equations that are generated. All three instances are generated in the execution file, CRITPATH.EXC.

9.6 Use of set maps.

Let us first review what a set map is (c.f., §7.1). It maps a domain, which is a product of sets, to a member of another set, called its range. In the model file, the syntax for a set map is:

name(*domain*) [*meaning*] = *range*

For example, suppose J is a set of journeys, and CTY is a set of cities. We can have the set maps,

```
ORIGIN(J) origin of J = CTY
DEST(J) destination of J = CTY
```

The first map says that for each journey, there is a city, called its origin; and, the second says there is a city called its destination.

A set map defined this way is an explicit set in that the data file specifies each member. This is in a list form that is different from a primitive set in that the first part of each member line is the domain member. The last entry in the line is the range set member. In our example, suppose J = {1, 2, 3} and CTY = {CHIC, SEAT, DENV}. Then, the following is a valid data specification for either ORIGIN or DEST.

```
1  CHIC
2  DENV
3  SEAT
```

Note that the meaning of each member does not appear. In forming syntax, the meaning of the range member is used as the meaning of the map member. Also, not every domain member needs to be present in the data file specification. The map value is *undefined* for each absent domain member in the list. If a domain member appears twice, an error message is given.

If the RANDOMIZE option is used to fill a set map, the domain members are randomly selected independently, then the range member is selected randomly using the same options. To ensure that it is complete (i.e., that every domain member is mapped to a range set member), set the RANGE option to a large enough value. If the domain sets have known cardinality, simply set the RANGE value to the product of these cardinalities (or greater).

Now let us consider a model where set maps are useful. Figure 9-31 lists a model, called JOURNEY, which has set maps, ORIGIN and DEST, as in the above example (to follow, READ MOD JOURNEY then READ DAT).

```
MODLER  ... LIST MODEL
                                          Model JOURNEY
                                       Aircraft Assignment
  Minimize COST
    Subject to:

   COST =  COST_AJ(A,J)*X(A,J)
   D(org,des) =  SUM[j IN J:ORIGIN(j)= org,DEST(j)= des| LOAD(A)*X(A,j)]
               >= DEMAND(org,des)
   L(A) =  X(A,J) <= 1.0
    Decision variables:
                               X = 0, 1
```

Figure 9-31. Listing a Model (JOURNEY) that Uses Set Maps

The set maps are used in the sum expression of the demand constraint. It is natural to define demands in terms of origin and destination, which is why the DEMAND table has that domain. The assignment variables, however, are naturally defined with journey in its domain. Also, two journeys could have the same origin and destination, so it would not be precise to use these cities as a definition of a journey. This requires a mapping from journeys to origin and destination cities, as shown. When the sum is instantiated, the only journeys that appear in the sum ($j \in J$) are those with the cities in the domain of the demand constraint.

Figure 9-32 displays the sets (in extended format), and we see the two set maps, ORIGIN(J) and DEST(J), whose members were read from the data file, JOURNEY.DAT.

```
MODLER  ... D S //E
  Set J has 4 members...journey
      = 1,...,NJ
  Set CTY has 4 members...city
      From JOURNEY.DAT
  Set A has 5 members...aircraft
      = 1,...,NA
  Set ORIGIN(J) has 4 members...origin of J
      From JOURNEY.DAT
  Set DEST(J) has 4 members...destination of J
      From JOURNEY.DAT
  Set org has 4 members...city
      = CTY
  Set des has 4 members...city
      = CTY
```

Figure 9-32. Displaying the Sets for the JOURNEY Model

Figure 9-33 lists these set maps. Each of the four journeys have an origin and a destination. For example, journey 1 flies from Chicago to Denver. You might write the syntax and matrix files for this instance. Then, solve the instance, and observe that the linear programming solution (with $X = 0, 1$ replaced by $0 \le X \le 1$) is fractional.

```
MODLER  ... L S ORIGIN,DEST
  Set ORIGIN(J) Source: JOURNEY.DAT
    J CTY city
    ============
    1 CHI Chicago
    2 DEN Denver
    3 MON Montreal
    4 DAL Dallas
  Set DEST(J) Source: JOURNEY.DAT
    J CTY city
    ============
    1 DEN Denver
    2 MON Montreal
    3 DAL Dallas
    4 CHI Chicago
```

Figure 9-33. Listing the Set Maps, ORIGIN and DEST

Chapter 10. Model Management

Good questions outrank easy answers.
Paul A. Samuelson

This chapter elaborates on some of the modeling process pertaining to model management. Managing a model includes maintaining documentation, conducting periodic reviews for purposes of verification and/or validation, and operating the model. Operations include running instances, called *scenarios*, in order to address questions for which the model was built in the first place. A collection of scenarios, related to address a common set of questions, is called a *case study*.

First, we give an overview of model management, in the context of using MODLER. In some cases, this requires a companion system, ANALYZE, which is designed to support analysis of one instance.

Second, we examine a case study, which is included with the MODLER library of models and data. Third, we bring attention to directories that were designed to aid model formulation.

10.1 Overview.

Figure 10-1 gives an overview of the components pertaining to model management. Using MODLER, the Model Description file is the model file, and the Model Documentation file is the explain file. Note that this is high-level documentation – that is, it is about the model description apart from its history of operations.

Base data could change over time, and there could be more than one such data file. Its contents are data values that generally apply to all instances, such as members of sets or number of time periods. (If sets do not change, one syntax file is sufficient to pass syntax information to ANALYZE.)

The model manager maintains a History of Operations file. This consists of summary information plus file specifications, usually cross referenced in the Case and Scenario Information file. The latter has detailed information about cases composed of scenarios designed to address a common set of questions, such as the following.

- When and where should new plants be constructed?
- Where should finished products be manufactured, and where should they be assembled?
- What is the impact of a change in policy on total production?

Each question becomes a case study that involves a collection of scenarios to deal with uncertainty in data values. In some cases, such as a change in policy, there can also be uncertainty in some of the model structures, such as the configuration of a distribution network that is part of the model.

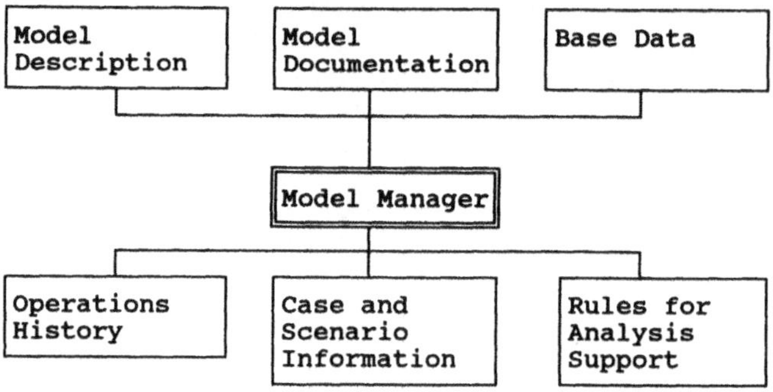

Figure 10-1. An Overview of Components for Model Management

The Case and Scenario Information file contains details that not only document each scenario and case results, but also gives file specifications for each instance – that is, the matrix file used as input to the optimizer (and to ANALYZE).

Part of model management is to provide support for the analysis of scenarios that comprise a case. One way to do this is by creating rule files for ANALYZE. Presently, this is done manually, as part of the model manager's responsibility. In the future, much of this will become automated, unburdening the human model manager from doing all that is required. Instead, the computer model manager will assume responsibility for generating and maintaining these files, and the human being will be responsible for checking the results and modifying the rules, as necessary, for automatic generation.

10.2 A case study.

The file WOODSCEN.DOC, which is included with MODLER data, describes a case study using the WOODNET model. The case involves uncertainty in supply and demand values. The scenarios are defined for the nine combinations of Low, Medium and High levels on both supplies and demands. The base case, often defined as the most likely state of nature in the face of uncertainty, is in this case the Medium×Medium scenario.

The file WOODSD.EXC is an execution file that generates a matrix file for each of the nine instances (one per scenario) plus the syntax file that applies to all scenarios. The base case is generated first, then each scenario is a modification of the base case by reading the associated data file. For example, if you examine WOODSD.EXC, you will see that each scenario is generated with the following specifications.

```
READ PAC WOODNET
READ DAT WOODsd
WRITE MAT WOODsd
```

where *s* = supply assumptions = L, M or H (for Low, Medium and High, respectively), and *d* = demand assumptions = L, M or H (also, for Low, Medium and High, respectively).

The first line reads the packed file for the base case, called WOODNET (equivalent to WOODMM in the syntax for naming scenarios). Then, the data file for the scenario is read into memory, overwriting the base case values. For example, if you look at WOODLM.DAT, you will see that only the supply tables are specified with new data, while the demand table remains the same as the base case (i.e., Medium level). Finally, the matrix file is written using the same naming convention for the scenarios. For example, the matrix file for the scenario of Low supply and Medium demand is named WOODLM.MAT.

After the matrix files are written for each scenario, the execution file (WOODSD.EXC) proceeds to write another execution file for use with ANALYZE. If you are familiar with ANALYZE, you can examine this.

To try all of this out, you can EXEC WOODSD in MODLER. Then, use MINOS to solve each scenario (9 instances). Finally, enter ANALYZE and use _SETUP to change your prefix for EXECUTE to where MODLER put the execution file it generated, called ANALWOOD.EXC. In ANALYZE, enter EXEC ANALWOOD. This will do two things. First, it reads each scenario's matrix and solution file, sets up the basis and saves it as a packed file. This is for further analysis of each scenario. Second, it writes a summary (using a generic rule file, called STATS). The collection of summaries are all in one file, called WOODSD.DOC.

When using MODLER plus other systems, like ANALYZE and MINOS, it is useful to be in a windows environment. In that case you can have four windows active, as shown in figure 10-2. One window is for MODLER; one is for an editor (in case you want to change the model file or a data file); one is for optimization (using an on-line version of MINOS); and one is for ANALYZE. The editor window shows ENDDATA, which is the last line of a model or data file; the other windows show system prompts. Depending upon your computer's memory, you might also have multiple ANALYZE windows to analyze results across scenarios, where each ANALYZE window contains a scenario.

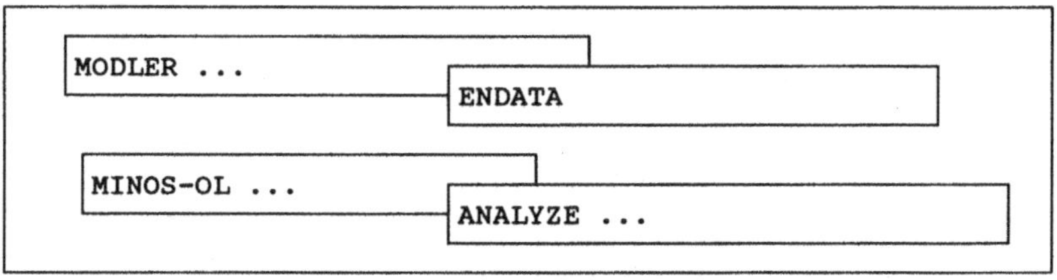

Figure 10-2. Using MODLER and ANALYZE in a Windows Environment (plus Editor and Optimizer)

10.3 Using directories to aid model formulation.

There are some documentation files, which came with MODLER, that begin with DIR. These are directories, which provide information about the MODLER library. To see a list of directories, enter ? DIR (using the MODLER dictionary, which you might expand if you add your own files, as described in Chapter 8).

Among the directories, you will find three that can help with formulation, namely:

DIRACTIV is a directory of decision objects in models.
DIRSETS is a directory of sets and related keywords in models.
DIRTABLE is a directory of tables in models.

where each directory file has suffix, .DOC (PRINT type in _SETUP). These refer to the model files that came with MODLER, which are listed in another directory file, DIRMODS.DOC.

Suppose, for example, you know you have a set of regions. You may want to know which of the MODLER model files also have a set of regions, so you can look at that model file for ideas on formulation of your linear program. You will see the following entry.

```
REGION............BLENDREF,BLENDRGN,LANDUSE,LOCATION,TRANSCAP,
                  TRANSGEN,TRANSLGL,TRANSTD,WOODNET
```

where REGION is a set/keyword. The notion of keyword in this context is that the set name need not literally be REGION. It could, for example, be called a location.

What follows is a list of models that are regional. For example, BLENDREF.MOD contains a regional blending model description. Other sets/keywords you will find are PROCESS, TIME, and MATERIAL. Examine this dictionary and the model files referenced for those sets/keywords you believe you will have in your model.

The directory of tables (DIRTABLE.DOC) is another cross reference that keys on table names, or equivalent keywords. If you examine this directory, you will find entries such as BUDGET, COST, DEMAND, DISTANCE, PROBABILITY, REVENUE and WASTE. For each entry, a list of model files is given, which you can examine to get ideas on your formulation.

Similarly, the directory of decision objects (DIRACTIV.DOC), which are mostly activities, give keys like ALLOCATE, CONVERT, HARVEST, SCHEDULE, and TRANSPORT. As before, a list of model files is given for each keyword.

Directories, such as these, offer a useful way to maintain a library of models and retrieve from it using keywords about data and/or decision objects. These directories can be expanded as new models are formulated and made accessible in a shared modeling environment.

Appendix A: Troubleshooting

This appendix describes error messages and how to use them for one aspect of model debugging. The print file, ERRORS.DOC, begins with the following.

```
Error messages are given when MODLER cannot recognize what you
mean, or if some necessary condition is violated.  Syntax errors
exemplify the former, and model completeness, such as including
at least one objective equation, exemplifies the latter.  In
some cases, these error messages do not immediately pinpoint the
error.  As in model debugging, one takes information and tries
to diagnose the cause.  Because error messages are imperfect in
the information they provide, this list of error message
explanations is not a complete guide to debugging.  Further,
MODLER is an imperfect language with some limitations and
quirks.
```

This file is your best starting point for understanding how to use the information from an error message.

Sometimes the error message is not a very accurate description of the cause. One such case is remembering that the lengths of names, like names of parameters, are limited to eight characters.

Besides syntax, another common cause of an error is the I/O ERROR message. This occurs if the file does not exist. Check your full file specification by using _SETUP to see the default prefixes and suffixes. You might also have inadvertently put an unprintable character in the name of the file, or in the NAME specification in the model file. If so, when you try to read data using this name as the default, you get an I/O error message. Try over-riding the default with a fully specified data file. If that works, you probably have an error in the name, in which case you need to go back to your editor and re-enter the name.

Appendix B: The MODLER Library

This appendix describes the MODLER library of model files with example data. A summary list is given in DIRFILES.DOC, and a brief description of model files is in DIRMODS.DOC. Most, but not all, of these were used in the User's Guide to illustrate MODLER's syntax and capabilities.

The MODLER library has three kinds of model files. One kind is *generic*, sometimes called a *template*, or *building block*, that one can find in most textbooks that give an introduction to linear programming. The following are recommended.

A. Charnes and W.W. Cooper, *Management Models and Industrial Applications of Linear Programming*, vols. I and II, Wiley, New York, NY, 1961.

H.G. Daellenbach and E.J. Bell, *User's Guide to Linear Programming*, Prentice-Hall, Englewood Cliffs, NJ, 1970.

G.B. Dantzig, *Linear Programming and Extensions*, Princeton University Press, Princeton, NJ, 1963.

J.S. Dyer and R.D. Shapiro, *Management Science/Operations Research Cases and Readings*, John Wiley & Sons, New York, NY, 1982.

G.D. Eppen and F.J. Gould, *Quantitative Concepts for Management*, Prentice-Hall, Englewood Cliffs, NJ, 1979.

F. Glover, D. Klingman and N.V. Phillips, *Network Models in Optimization and Their Applications in Practice*, Wiley, New York, NY, 1992.

H.M. Wagner, *Principles of Operations Research*, Prentice-Hall, Englewood Cliffs, NJ, 1969.

The following model files are generic.

ASSIGN .... Assignment problem
BLEND...................... Simple blending problem
BLENDGEN .... General blending problem
BLENDRGN .... Regional blending problem
BLENDRT .... Regional, dynamic blending problem
GENET..................... Generalized min-cost network flow
INVENTRY .... Inventory control
JOURNEY .... Aircraft assignment
MANPOWER......... Manpower planning
MULTIPLE .... Multiple-objective network flow
NETFORM .... Netform
POOLING .... Pooling (linearized)

PRODIST................Product distribution
PRODMIX..............Product mix
RECRUIT...............Recruitment of personnel
SCHEDULE...........Scheduling jobs
SUPDEMEQ..........Supply-Demand Equilibrium model (step functions)
TABLEAU..............Tableau form
TRANSCAP............Capacitated transportation problem
TRANSGEN...........Generalized transportation problem
TRANSLGL............TRANSGEN with logical constraints and binary variables
TRANSTD..............Standard transportation problem

The second type of model file is also generic, aimed at giving an alternative representation. Examples are the inventory control models. In one representation, we used control tables with simple, implied summations. In another representation, we used reference forms. Besides using them in this User's Guide, the installation test (ATEST.EXC) reads them to check that MODLER works in your computing environment.

The following model files are alternative representations of generic models.

BLENDALT...........Alternative BLENDGEN
BLENDREF...........Alternative BLENDGEN
INVCTRL...............Alternative INVENTRY
INVREF..................Alternative INVENTRY
SUMTEST1............Alternative TRANSGEN
SUMTEST2............Alternative TRANSGEN
SUMTEST3............Alternative INVENTRY
SUMTEST4............Alternative INVENTRY

The third type of model file is one taken from the literature. In each case, the model file cites the reference, which you can print. The following model files are of this type, which are classified by application area (authors given).

Agriculture
LANDUSE..............Land allocation (Heady & Hall)

Construction
CRITPATH.............Critical path scheduling (Stark & Mayer)
EARTHMOV.........Minimize earth moving costs (Stark & Mayer)

Database
FILEMERG............File merger (Glover & Klingman)

Environment

WATERPOL .... Water pollution control (Graves, Hatfield & Whinston)

Manufacturing

DYNLOTS .... Dynamic economic lot size problem (Wagner & Whiten)

Urban planning

LOCATION .... Facility location for solid waste (Marks & Liebman)

Besides these model types, there are some miscellaneous model files, which you can examine. See DIRMODS.DOC for a complete directory of model files. Further, see other DIR*.DOC files that contain cross references to model objects. For example, DIRSETS.DOC is a list of set types, like regions, materials and time periods, that cite which model files contain the types. This is useful to assist model formulation by seeing representations similar to your own problem.

> It is likely the MODLER library will expand as new versions are released, and you are encouraged to notify the author about any models you can contribute.

Index

GPSR Compliance

The European Union's (EU) General Product Safety Regulation (GPSR) is a set of rules that requires consumer products to be safe and our obligations to ensure this.

If you have any concerns about our products, you can contact us on ProductSafety@springernature.com

In case Publisher is established outside the EU, the EU authorized representative is:

Springer Nature Customer Service Center GmbH
Europaplatz 3
69115 Heidelberg, Germany

Batch number: 10761599

Printed by Printforce, the Netherlands